# PREVENTING WASTE
# at the
# SOURCE

## Norman J. Crampton
*Indiana State University*

CRC Press
Taylor & Francis Group
Boca Raton  London  New York

CRC Press is an imprint of the
Taylor & Francis Group, an **informa** business

CRC Press
Taylor & Francis Group
6000 Broken Sound Parkway NW, Suite 300
Boca Raton, FL 33487-2742

© 1999 by Taylor & Francis Group, LLC
CRC Press is an imprint of Taylor & Francis Group, an Informa business

First issued in paperback 2019

No claim to original U.S. Government works

ISBN-13: 978-0-367-44772-4 (pbk)
ISBN-13: 978-1-56670-317-8 (hbk)

Visit the Taylor & Francis Web site at
http://www.taylorandfrancis.com

and the CRC Press Web site at
http://www.crcpress.com

Library of Congress Card Number

Library of Congress Cataloging-in-Publication Data

Crampton, Norman.
    Preventing waste at the source / Norman J. Crampton.
        p.   cm.
      Includes index.
      ISBN 1-56670-317-4
      1. Factory and trade waste—Management.   2. Waste minimization.
    I. Title.
    TD897.5.C73 1998
  658.5'67—dc21                                                          98-29657
                                                                            CIP

## About the Author

Norman Crampton is director of the Indiana Institute on Recycling, located on the campus of Indiana State University, in Terre Haute. The Institute provides a technical information and consulting service on waste prevention and recycling, serving a clientele including businesses, municipalities, and Indiana residents.

Since 1995, with major support of the WasteWi$e program of the U.S. Environmental Protection Agency, the Institute has been developing a reference center of case studies on business and industrial waste prevention. Readers are invited to consult this resource on the World Wide Web at web.indstate.edu/recycle.

Mr. Crampton is the author of several books, including the first popular book about recycling, *Complete Trash: The Best Way to Get Rid of Practically Everything Around the House*, published in 1989. Before establishing the Institute office in 1990 he was secretary of the Institute for Solid Wastes, American Public Works Association.

# Contents

# Introduction

The basic idea of this book can be summed up in one sentence: Waste prevention will yield a much higher rate of return on investment than most other investments your company is now making.

Does that sound like a rash claim? It's not. In the chapters that follow, we'll take a look inside companies all across the United States where comparatively simple and inexpensive waste prevention initiatives have cut operating costs by thousands of dollars. These are not merely windfall savings but calculated, permanent reductions in production expense that continue to yield benefits year after year.

The surprising thing about these savings is that they usually include *unexpected* benefits. For example, when Acme Company sees that it no longer needs to use Shipping Crate A, the total cost saving will include several parts:

1. Avoided cost to purchase Crate A.
2. Reduction in Warehouse Space B to store Crate A.
3. Reduction in Clerical Time C to keep all the records associated with buying, storing, using, and replacing Crate A.
4. Reduction in Disposal Cost D or the final disposition of Crate A.

That's not all. By doing away with Crate A, Acme Company may also

5. Reduce Worker Injury Cost E stemming from the handling of Crate A.
6. Improve Throughput F and Turnover G by substituting more efficient handling of material than was possible with Crate A.

It's quite likely there will be an additional, significant benefit. By consuming fewer trees and using less power to turn trees into crates, Acme Company reduces its demand on the environment. As this list of benefits illustrates, waste prevention—using less stuff—usually entails *simple* changes in production routine, but it almost always pays *compound* interest.

## ASK WHY FIVE TIMES

Though opportunities for preventing waste are often simple and, in hindsight, quite obvious, they are usually camouflaged by years of company routine. "Why do we use Crate A? Because we always have!" One way to peel off layers of company practice is to adopt the Japanese method of inquiry: To discover the real reason for practically anything, ask *why* five times.

Here's an illustration. Suppose that you were puzzled by the spacing of railroad tracks in the United States—4 feet 8-1/2 inches apart. At first glance, that precise dimension looks like a deliberate choice based on efficiency and cost. Let's ask *why* five times:

1. *Why are U.S. railroad tracks spaced 4 feet 8-1/2 inches apart?*
   Because the U.S. adopted the standard British spacing of railroad tracks.
2. *Why are British railroad tracks spaced that way?*
   Because the dimension was adopted from the design of tramways that preceded railroads.
3. *Why were British tramways spaced that way?*
   Because the jigs used to build tramways were the same as jigs for horse-drawn wagons.
4. *Why did the jigs for wagons set wheels 4 feet 8-1/2 inches apart?*
   Because that was the distance between wheel ruts on ancient British roads, built by Imperial Rome for chariots.
5. *OK, why were the wheels of Roman chariots spaced 4 feet 8-1/2 inches?*
   Because that's the optimum width to accommodate two horses in harness, butt to butt!

The story may be part fancy but the point is not—specifications are not always as scientific as they seem. Ask *why* five times about Shipping Crate A and you may be led to an equally implausible beginning as the spacing of railroad tracks in the U.S. In fact, before deregulation of the trucking industry, packaging standards for motor freight were heavily influenced by the corrugated cardboard industry. And those standards were adopted largely from railroad freight regulations which, because railroads owned a lot of forest land along their rights-of-way, strongly favored crates made of wood. It's easy to see how established practice—"the way we've always done it"—can survive for years without serious question.

## MANY TARGETS OF OPPORTUNITY

Opportunities for waste prevention are many more than you might guess, even in the best run companies. In the next chapter we'll talk about how to examine a company, department by department, to identify wasteful *procedures*. Notice the emphasis on process: we'll be looking at everyday work routines, "the way things are done," as harbors of waste and unnecessary expense. To preview what we will find, here's a quick look at the largest targets for waste prevention in most business enterprises:

- Front office—*Paper* in all its forms, especially copy paper.
- Manufacturing department—*Production scrap,* all the leftover metal, plastic, wood, or fiber that remains after products are finished.
- Receiving/shipping—*Containers,* all kinds, both incoming and outgoing, both logistical and retail, with a lot of attention to corrugated boxes and wooden crates.
- Food service—All *single-use* items, such as throwaway paper and plastic.
- Housekeeping—Again, single-use items, such as trashcan liners that are replaced daily even when little if any trash has accumulated.
- Warehouse—*Stretch wrap,* another prime example of single-use materials. Many companies have found ways to curtail or eliminate stretch wrap from inventory.

## WASTE PREVENTION—A TOUGH SELL

Though examples of waste may seem obvious, and the potential savings may seem significant, getting management's approval to invest time and money in waste prevention can be a very tough sell. Understandably, it's routine for managers to expect some pretty solid evidence in advance of any commitment of capital: How much money are we going to spend? How quickly will we recover the investment?

If funds are requested for familiar kinds of things, such as a production tool identical to others that are reliably stamping out products on the manufacturing floor, the process is fast-track: Purchasing Machine X will increase Output Y and, everything else remaining equal, Gross Revenue Z. Familiar investments pass inspection much more easily than investments of either money or time in the comparatively unknown territory of waste prevention. How closely, for example, do companies scrutinize proposals to invest in information technology—computers and related items? It has been estimated that no more than one company in five has a process in place to cost-justify an investment in IT. Decisions are based on familiarity and faith—remarkable when you consider that IT investments are approaching $1 trillion a year!

By comparison, preventing waste at its source requires only small fractions of capital, sometimes none at all. Yet it can be a tough proposal to defend. Perhaps the reason is that it is hard to imagine the end result if the end is *less or nothing*. The very process of reduction sends shudders through many business enterprises because cutbacks are associated with failure and contrary to the natural business momentum of growth and increase. But the kind of reduction we are talking about is totally different. It is aimed at *preserving* profits and *sustaining* growth. "The business of source reduction is simple," writes Chaz Miller, a national authority on the topic. "A well-managed company will strive to reduce waste so that it can use its financial resources more efficiently."

## TRUST NURTURES THE PROCESS

Waste prevention thrives in a climate of trust—trust between companies and their suppliers and customers. Easy to say, very hard to put into play. In an intensely competitive world economy, a company's natural inclination is to keep its guard up, to defend against undependable suppliers on one side and fickle customers on the other. Nevertheless, businesses are moving rapidly toward more interdependence. There is no better evidence than the rapid spread of just-in-time delivery of materials in the auto industry. Example: Car seats manufactured at noon at Setex Co. in St. Marys, Ohio, are installed 4 hours later and 25 miles away at the Honda plant in Lima. Honda trusts Setex to deliver on a very short schedule. To comply, Setex has ruthlessly eliminated waste in production and shipping, including the total elimination of protective wrapping of car seats for transport to the auto plant. They are bolted to a reusable platform, rolled into a semitrailer truck, and delivered direct to the assembly line.

An even tighter bond between supplier and customer is working well at the Volkswagen truck and bus plant 150 kilometers south of Rio de Janeiro. There, suppliers to the plant are not merely close to VW, they're *inside* the plant, actually installing modules along the production line. From VW's point of view, the effect on waste prevention is dramatic. Incoming shipping containers—pallets, crates, racks, boxes—are drastically reduced. Such a close connection between supplier and customer may not be appropriate in every setting, but it illustrates what is possible.

## THE ENVIRONMENTAL DIVIDEND—DOING GOOD WHILE DOING WELL

This is a very practical book. We'll be pricing and measuring many things, and comparing alternatives, always with an eye to reducing cost by eliminating waste. In our approach, waste prevention and cost reduction will always go hand in hand. And that leads to the extra dividend of the entire process: While we are concentrating on rooting out waste, with the objective of reducing cost, simultaneously and unavoidably we'll be reducing the amount of stuff that goes to final disposal in a landfill or an incinerator. Disposal in a landfill carries the long-term risk of polluting groundwater, the source of drinking water for the majority of Americans. Disposal by burning in an incinerator immediately produces carbon dioxide, one of the prime suspects in the search for causes of global warming. Thus, by reducing the quantities of materials that ultimately are thrown away, companies exercise their corporate responsibility to preserving a healthy environment. In other words, by doing well for themselves—lowering costs—companies do good for their community. It's a compelling combination.

# CHAPTER 1

# How to Think About Waste Prevention

Suppose that you and a colleague decide to take a fresh look at the office and production area. You want to spot opportunities to reduce costs by preventing waste. But what exactly will you be looking for? And how will you know when you see it?

If you are already preventing waste at the source, you may find no trace of your wise action. Effective waste prevention leaves only an *absence* of something that once seemed indispensable. Unlike recycling, which looks like boxes, bales, and carloads of reusable scrap, waste prevention looks like nothing. It can be a beautiful sight!

The most important question to ask yourself, and others, as you look around the business for wasteful practices is, Why? Why do we do it this way? Why do we send out all these reports? Why is the dumpster full of pallets and cardboard boxes? Why are we producing so much scrap metal (scrap wood, plastic, paper, fiber, etc.) when these commodities are costly as production feedstock but earn so little as recycled scrap?

## WASTE PREVENTION IS NOT RECYCLING

The distinction between waste prevention and recycling is important to keep in mind when you examine company operations. You may already be doing a very good job of *recycling*. You may have systems in place to capture such things as used cardboard, office paper, production scrap, newspaper—even soft drink containers—and to send all these things back to the mill for processing and remanufacturing into new products. That is a good way to manage materials, but it is not the best way. Some years ago, the U.S. Environmental Protection Agency created a "hierarchy," a good-better-best ranking for the management of materials that would otherwise become municipal solid waste, or MSW. Here's the hierarchy:

### Solid Waste Hierarchy—The Three R's

1. REDUCE—Use less stuff, starting at the very front end of operations.
2. REUSE—Recover used materials intact, send them around the cycle again.

3. RECYCLE—Keep used materials out of the disposal system, send them to pro-
cessors as feedstock for manufacturing new products

From the bottom up—recycling is good, reusing is better, reducing is best. For
obvious reasons, it makes much more sense to curtail or eliminate unnecessary
materials "from the top," from the beginning of operations, rather than to incur
multiple costs—procurement, storage, inventory, handling, disposal, etc.—associated
with managing materials all the way through the production pipeline and dealing
with them creatively only when they reach the end of the pipe.


## KEY DISTINCTIONS

Let's note that *reduction* means the same as the two-word, technical term *source
reduction*, but that in this book we'll prefer another term we believe is more descrip-
tive and familiar: *waste prevention*. (Do not substitute the term *pollution prevention*,
which refers to minimizing the use and production of hazardous substances.) One
more definition is essential to describe the general class of materials we are concerned with: *municipal solid waste*, or *MSW*, refers to nontoxic discards from offices,
stores, factories, households, and institutions that are disposed of in landfills or
incinerators. MSW is sometimes called *refuse*. It is distinguished from *hazardous*
or *toxic* waste, both of which require special handling and disposal in special
facilities.

### CASE STUDY: Schumacher Electric Corp.

Rensselaer, Indiana

*"Never buy anything you're throwing away."*

Schumacher Electric, located in Rensselaer, Indiana, designs and manufactures
electrical transformers for power, audio, and other applications. The products are
small but heavy and require very secure packaging in corrugated cardboard cartons.
To prevent damage during shipment to customers, the transformers are cushioned
on the top, bottom, and sides with strips and rectangles of corrugated cardboard cut
to special dimensions.

Loren Snow, the plant manager, has a management maxim: "Never buy anything
you're throwing away." One day as he was walking around the plant, Snow observed
that the Shipping Department was receiving and warehousing a steady supply of
special-cut pieces of corrugated cardboard for use in packing finished products,
while in another part of the department there was always a large stack of old
corrugated cartons—discarded incoming packaging materials gathered up from here
and there around the plant. Standard procedure was to bale the old cartons for sale
to a paper recycler, about 16 tons of material a year.

For Snow, this was an "Aha!" moment. Why not convert the old cartons into
cushioning strips for customer shipments? He found a bandsaw in a corner of the

**Figure 1**  HOME MADE. Simple strips of corrugated cardboard packaging were costing a transformer manufacturer $60,000 a year to buy new. Now they are produced quickly as needed from incoming cartons, cut to size on an old bandsaw.

plant ("Every factory has at least one of these standing around," he said), moved it to the Shipping Department, and trained the receiving-shipping clerk to cut the old boxes into new packing strips. Under the new procedure, only torn or dirty old cartons go to the baler. Reusable cartons—90% of all incoming cartons—are flattened and stacked in neat piles. As packing strips are needed, the worker lifts a short stack of cartons to the bandsaw table and, guided by jigs, quickly cuts the various cushions, shelves, and partitions required to package transformers for shipment.

## Payback

During the last full year that Schumacher purchased corrugated packaging parts from outside suppliers, the total cost was about $60,000. Because the conversion to inside manufacture of packaging required no capital outlay (feedstock, equipment, space, and labor all were available), there was no payback period. The $60,000 reduction in annual operating expense was available immediately.

Beyond the obvious reduction in direct costs, Snow believes there may be significant additional savings when products that have been purchased outside are manufactured in-house from "waste." For example, he gained about 2,000 square feet of storage space that had been dedicated to the inventory of custom-made cardboard pieces from outside suppliers. "You also avoid all kinds of paperwork," he observes. There is no longer any need to maintain parts numbers, for example; no need to generate purchase orders, maintain inventory records, and handle all the paperwork associated with paying vendors. Although he has not calculated all these savings, Snow believes the reduced cost of clerical and supervisory time easily offsets whatever additional expense might be incurred from producing a necessary packaging product in house—from scrap.


## SEVEN BASIC APPROACHES TO PREVENTING WASTE

Waste prevention focuses on the way we use materials. Practically all the strategies for preventing waste can be organized under the following seven approaches to materials:

1. Eliminate.
2. Reduce weight or thickness.
3. Increase capacity.
4. Replace single-use with multiple-use.
5. Purchase for long life.
6. Redesign.
7. Transform waste into product.

There are countless variations and combinations of these seven basic themes. At Schumacher Electric (cited above), plant manager Loren Snow saw an opportunity to eliminate a purchase (Basic Approach No. 1) by transforming a waste into a product (7), illustrating the principle of reusing materials instead of tossing them after a single use (4). The following seven cases have been selected to illustrate primarily one of the seven basic approaches to waste prevention, but in each you will probably see connections to more than one approach.

### CASE STUDY: Alpine Windows

Bothel, Washington

*Eliminating a crate cuts annual costs $265,000.*

Alpine Windows is one of the largest window manufacturers in the Northwest, employing 400 people. Previously, Alpine's glass supplier, a California company, delivered bulk packages of window glass in wooden crates. For example, a number of sheets of glass measuring 96 × 120 inches would be packed upright inside a wooden crate, with steel strapping on the exterior to secure this shipping package. Crateloads weighed 4,000 pounds and were shipped by common carrier on a flatbed truck. The glass supplier charged $50 each for the 12 to 24 crates required each day for shipments to Alpine. Thus, daily crating costs ranged from $600 to $1,200.

Because of the distance between glass supplier and window factory—900 miles—and because of the use of common carrier rather than a truck dedicated to this delivery service, Alpine did not have the option of deadheading empty crates back for reuse. So the empties became an expensive pile of scrap wood that Alpine

**Figure 2**   OUT OF THE BOX. Huge sheets of glass used to arrive at a window factory in heavy wooden crates. Now they are shipped uncrated, leaning against an A-frame on a flatbed truck.

**Figure 3**   TROLLEY RIDER. Packs of glass are lifted off the delivery trailer by a forklift rigged with a special boom, then rolled inside on a trolley. Eliminating crates saves a window maker $265,000 a year.

placed out for scavengers to haul away free. Plenty of eager takers descended on the free woodpile. Periodically, Alpine gathered up the broken remnants and hauled them to a wood recycler at an average cost to Alpine of $200 per trip. All scrap wood was put to good use in one way or another, but it was an expensive community service!

The opportunity to eliminate wooden crates became apparent when Alpine adopted new technology for feeding individual sheets of glass into the glass cutter. Previously, whole crates were removed from the flatbed truck by forklift and placed on a heavy-duty cart for transport to the production area, where the crate was opened and sheets of glass were picked up one at a time by suction cups and placed on the feeder table.

The new feeder method at Alpine enables the glass supplier to lean uncrated packs of glass against an A-frame structure on the flatbed truck. Packs are covered with tarpaulins and securely tied down. For delivery at Alpine, glass packs are unloaded by a forklift rigged with a boom and webbed nylon slings. Packs are placed on a trolley rigged with an A-frame and moved to the production area, where individual sheets of glass are peeled away from the pack and permitted to fall onto the feeder table. Air pressure slows the rate of fall, preventing breakage.

### Payback

Alpine invested about $400,000 in the new free-fall system of placing glass on the feeder table. The equipment has a service life of 20 years. The new handling system also reduces restock time, because uncrated packs of glass contain 50% more sheets than crated packs. Alpine estimates the labor savings at $35,000 a year. The combination of reduced crating costs and restock time enabled Alpine to recover its capital investment in about 20 months.

### CASE STUDY: McDonald's

Oak Brook, Illinois

**Reducing** the hash browns bag saves $4 million.

McDonald's has become widely known for its initiatives in waste prevention and solid waste reduction. Between 1991 and 1996, 105 various projects under the McDonald's/Environmental Defense Fund Waste Reduction Action Plan (WRAP) yielded very significant savings in the use of materials.

When making decisions about how to package its food products, the company weighs four factors:

1. Environmental impact
2. Cost
3. Customer satisfaction
4. Effect on store operations

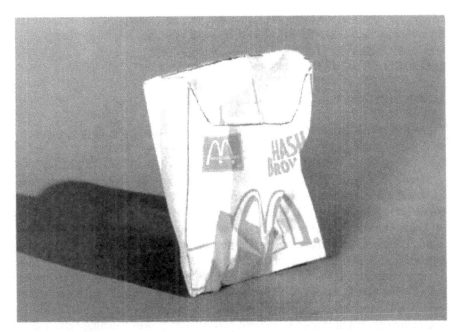

**Figure 4** SLIMMED DOWN. By switching to a lightweight paper envelope for its hash brown potatoes, McDonald's saves $4 million a year in materials, shipping, storage, and disposal.

For many years McDonald's had packaged its hash brown potatoes, a breakfast menu item, in a cardboard-like carton with a paper basis weight of 110 pounds. The container worked well. Mainly it was thick enough to keep grease from soaking through to customers' hands. Changing to a lighter weight paper seemed out of the question because this essential quality would be diminished or lost. But a development in paper coating technology made it worthwhile to experiment with the hash browns container. 3M Company came out with a coating called Scotchban FC-807, designed as a food-safe grease barrier. This enabled McDonald's to design a new container made of white paper, basis weight 28 pounds, or about 75% lighter than the cardboard container.

*Payback*

The new generation of hash browns bag is printed with the familiar arched M and other simple graphics. It weighs one-quarter the weight of the old cardboard container, and it requires much less energy to produce. The lighter container has reduced the trash load: McDonald's calculates that, worldwide, the new bag has reduced the company's production of solid waste by about 3.5 million pounds a year. Additional savings accrue from the reduction in space required to ship bulk quantities of the hash browns bag to McDonald's stores. Bulk packages of the bag occupy only 30% of the space (cube) of comparable quantities of the old carton. McDonald's believes its total savings, including the reduced cost of a lighter bag, reduced cost of disposal, and reduced shipping and storage space approximate $4 million a year.

## CASE STUDY: Madison Chemical Co., Inc.

Madison, Indiana

***Increasing the capacity*** *of bulk shipping containers from 55 gallons to 350 gallons reduces costs $19,000 a year.*

Madison Chemical is a medium-sized formulator of nonhydrocarbon-based industrial chemicals used for conversion coatings, cleaners, lubricants, rust preventatives, and paint strippers. Previously the company stored and shipped its product in 55-gallon drums, using some 3,000 drums a year. Empty drums were delivered to a reconditioning contractor for cleaning and testing at $10 per drum before refilling by Madison. Drums could run this cycle up to four times before they were scrapped.

After studying the advantages, Madison converted to the use of 350-gallon intermediate bulk containers (IBCs) to reduce the number and overall cost of containers, and to respond to customer concerns over drum handling and disposal. The IBC is constructed of stainless steel and plastic. With proper handling and cleaning it can be used indefinitely.

Madison found that replacing 55-gallon drums ($20 each) with 350-gallon IBCs ($500 each) saved about $19,000 a year in overall cost, even including the higher $30 per drum cost to recondition the IBC. These savings are calculated on the basis of purchasing 100 IBCs per year. As the company adds to its inventory of IBCs, savings will increase.

Madison also implemented a system for certain large customers wherein it replaced large numbers of circulating 55-gallon drums with IBCs that remain permanently at the customer's site and are refilled from bulk tanker trucks. The IBCs do not require cleaning because they are refilled with the same chemical. The IBCs used in this service are specially configured to include detachable, four-way-entry, reusable skids manufactured of plastic.

### Payback

Madison Chemical estimates it earned back the investment in IBC containers for shipping alone (not including storage) in 6 months. And there were additional benefits:

1. Reduced handling of product in storage and no shipping of raw material containers.
2. Reduced handling of finished product in IBCs for shipment to customers versus multiple barrels.
3. Easier and safer stacking of IBCs compared to barrels.
4. Better customer relations—customers said they liked the IBCs because of ease of handling and the ability to meter material usage.
5. Reduced product residue in empty IBCs because the discharge valve is located on the bottom.

**Figure 5**   CINCHED. For warehousing operations, large rubber bands work as well as stretch-wrap—and can be used repeatedly. Making the switch, a Wisconsin factory saves nearly $50,000 a year.

### CASE STUDY: Traex/Menasha Corp.

Dane, Wisconsin

*Replacing single-use stretch-wrap with multiple-use rubber pallet bands saves $49,706 a year in warehouse operations.*

Traex manufactures injection molded plastics and other products for the food service industry. Finished products are boxed in corrugated containers and stored on pallets in the warehouse until needed to fill orders. Previously the company stretch-wrapped the pallet-loads by hand for transfer to the warehouse. The wrapping process was time-consuming and required most of one employee's attention over three shifts. The process also exposed workers to the risk of back injury from stooping and straining.

Traex was motivated to make a change in this method of warehousing when its waste hauler said it would begin requiring customers to bale discarded stretch-wrap. Purchasing a baler would be costly and require floor space and labor. Traex began testing rubber pallet bands as a replacement for stretch-wrap. The unstretched bands measure 3/4 inches wide by 92 inches in circumference, adequate to span a load of cartons on a standard 40 × 48-inch pallet. Typically, two bands are used per pallet, one looping around the top layer of cartons and the other at about mid-load. Bands are applied by hand, an operation that takes one worker about 10 seconds per band. (By comparison, stretch-wrapping required about 1 minute per pallet.) When pallets are brought from the warehouse to fill orders, the bands are quickly and easily snapped off the load and tossed in a small barrel for reuse. At the Traex plant, it takes a pallet band about 6 weeks to make the round-trip from initial placement around a pallet-load at the production line, to warehouse, back to the order fulfillment department where bulk loads are broken down as orders are filled.

With training provided by the warehouse manager, this change in operations was adopted without difficulty. Employees like the rubber bands better than shrink-wrap. During a 10-month test period, only four bands broke. Apparently there is no risk to workers from snapping bands.

## Payback

The most important saving is reduced labor cost. Traex calculates that it recovered the cost of its initial supply of rubber bands in just four three-shift production days. Here is the company's summary of overall annual savings:

| | |
|---|---|
| Reduced labor | $50,112 |
| Reduced disposal expense | 400 |
| Reduced purchase of stretch-wrap | 195 |
| SUBTOTAL | $50,707 |
| Less 550 rubber bands @ 1.82 | − 1,001 |
| NET SAVING | $49,706 |

Reduced demand for stretch wrap has freed some storage space, and additional labor has been saved by avoiding the baling of stretch-wrap. Occasional use of temporary workers for unwrapping stretch-wrap has ended, as have occasional injuries to workers from using a knife to cut away stretch-wrap.

## CASE STUDIES: Minnesota

### Grand Rapids and Minneapolis

**Purchasing long-life** fluorescent bulbs to replace incandescent bulbs cuts electricity, waste disposal, and labor costs.

The Minnesota Office of Environmental Assistance (OEA) was one of the first state agencies to reach out to private enterprise with a waste prevention program. To demonstrate a variety of approaches, the OEA has assisted a variety of businesses in identifying targets for waste preventive action and has documented the results.

One of the opportunities at the Itasca Medical Center was conversion of exit sign lights from incandescent to fluorescent bulbs. There are 18 exit signs throughout the facility. All were replaced, with the following results: *Waste stream reduction for this item, 89%.* Exit signs are illuminated continuously. The incandescent bulbs were rated at 2,500 hours of service life but lasted somewhat less, with 18 signs requiring a total of 67 new bulbs a year. Fluorescents required 7.2 replacements per year.

Frattalones Ace Hardware, Minneapolis, converted from incandescent bulbs to 25-year-life LED bulbs for exit signs. Two 20-watt bulbs, using a total of 40 watts per sign, were replaced with LED bulbs using a total of 2 watts per sign. Payback for the labor and conversion kit occurred within 9 months. Thereafter, cost savings were 97 percent ($34) per sign per year. Waste volume reduction was 95 percent (127 cubic inches) per sign per year. Waste weight reduction was 96 percent (2.3 pounds) per sign per year.

### CASE STUDY: Haworth, Inc.

Holland, Michigan

**Redesigning** the traditional printed catalog as a CD-ROM vastly reduces paper consumption and improves presentations to customers.

Haworth, Inc. is the second largest manufacturer of office furniture in the U.S. Like many companies selling merchandise to the business sector, Haworth has for years printed a large, frequently updated catalog, along with numerous related publications and forms. In 1992, the company began to develop an alternative, paperless way to distribute this information—SourceBook—a CD/ROM that runs on Windows 95.

**Figure 6**   COMPACTED. Haworth replaces the old-style printed catalog with a CD-ROM, a costly project but a likely source of long-term printing savings and sales productivity increases.

This has been an expensive project. Start-up costs have included $500,000 for software development; and $2 to $3 million more to train 600 people, equip 260 market managers with laptop computers, and other costs. Despite the heavy initial expense, Haworth believes it is making a wise investment that will pay off in the long term. The company says it has already realized these benefits:

1. *Decreased printing costs and paper usage.* Because of frequent price and product changes, Haworth over the years has consumed vast amounts of paper, and spent more than $1 million annually for updated sales literature. Introduction of the CD will enable the company to reduce its annual printing budget by 40% by 2005.
2. *Higher quality client presentations.* With the CD, salespeople custom-design client presentations on the computer. The old way was to cut and paste from catalogs, which wasted large amounts of paper and yielded an inferior presentation by comparison. Response time to customer questions also has been reduced, since all essential information is loaded on the CD/ROM. (Haworth estimates that the 2,000-plus documents in SourceBook equal a stack of paper 6 feet tall. An updated CD is sent out monthly.)
3. *New training opportunities.* SourceBook helps to train employees and customers. For example, one video clip on the CD explains how wood veneer furniture is made and shows the various kinds of wood available.

### Payback

Systems Analyst Ted Evans says, "We believe that from a marketing perspective alone, SourceBook has already paid for itself through customer satisfaction, the general impressiveness and high visibility of the product, and the increase in employee efficiency."

### CASE STUDY: Haarmann & Reimer

Elkhart, Indiana

*Transforming a production waste into cattle feed saves $584,000 a year.*

Haarmann & Reimer (H&R) is a major producer of citric acid used in the production of such items as fruit flavored drinks, sodas, and orange juice supplement. In the early 1990s, H&R's parent company, Bayer Corporation, initiated a company-wide waste reduction program known as WRAM—Waste Reduction and Management. Mycelium landfill waste at H&R was soon identified as the fourth largest waste stream among Bayer's entire enterprise of more than 50 manufacturing sites.

Mycelium is the residual husk of selected enzymes added to the citric acid production process to enhance biologic fermentation. Because of the low weight and "blowing" properties of mycelium, H&R was paying a premium to haul the waste and dispose of it in a landfill.

A major alternative to disposal appeared to be the use of mycelium as a supplement to cattle feed. In 1994, Bayer commissioned Purdue University to study the nutritional value of mycelium as cattle feed. The study concluded that 1 pound of dry mycelium was the nutritional equivalent of 1 pound of corn, assuming that mycelium did not exceed 30% of the total diet.

Other alternatives to disposal also were evaluated, including land application of mycelium as a soil nutrient; use of mycelium to remove small particle solids from wastewater in mining operations; composting; and decomposition of mycelium in anaerobic waste treatment reactors.

Because of the comparatively low price of feed corn at the time of its investigation, H&R had difficulty finding local buyers of mycelium as feed. But when feed corn prices rose, a major farm in lower Michigan expressed interest, agreeing to take waste mycelium free but to pay shipping costs. To increase solids from 20% to 40% dry basis, H&R installed a filter press. This cut the shipping weight in half, reduced shipping costs, and doubled the daily amount of wet feed that cattle could consume.

### Payback

Bayer/H&R invested more than $1 million in the study performed by Purdue University, as well as new equipment costs. But since implementation of the new program, H&R has totally eliminated some 10,000 tons per year of mycelium from its waste stream. Out of projected annual savings of $584,000, avoided landfill tipping fees alone total $221,000. A payback period of less than 2 years is expected.

## THE POWER OF COMPOUND INTEREST

Financial planners like to point out how putting aside a little money on a regular basis over a long period of time can produce a quite sizeable nest egg. The combination of compound interest plus the passage of time turns small savings into large savings. We see exactly the same thing happening in each of the preceding case studies of waste prevention. At McDonald's, for example, the small amount of money saved per unit by shifting from a cardboard container to a lightweight paper envelope for hash brown potatoes is multiplied—compounded—by millions of units per day and hundreds of millions over a year. One small step in waste prevention yields a huge, permanent reduction in waste and operating cost.

Simple initiatives are much more common than complex projects to prevent waste, and they can yield just as much beneficial effect. At Traex/Menasha, switching from stretch wrap to rubber bands for containing pallet-loads was as easy as asking around for ideas among supplier companies, and experimenting in the warehouse. At Madison Chemical, the economies of moving chemicals in 350-gallon drums seemed obvious from the start; the only thing missing was some assurance of customer acceptance. By comparison, Haworth's conversion from the familiar printed catalog to a CD-ROM took a great deal more imagination, time, and money. And the payback is expected to be sizeable, as well.

If the dividends of waste prevention are so substantial, why does it take some companies so long to start cashing in? The answer is familiarity and habit. It is just as hard for companies to discard the familiar, habitual way of doing things as it is for individuals. Yet it's amazing how quickly companies capitalize on waste prevention opportunities once they sense the cost implications. Nancy Hirshberg, director of natural resources at Stonyfield Farm, summarizes the idea: "Waste in any form is lost profits. Maximizing efficiency through reducing our waste stream is not only good for the environment—it's good for our bottom line as well."

In the next chapter we'll look at the tactics companies use to position themselves at the front line of waste prevention.

CHAPTER **2**

# Organizing for Action

If you want to find where the waste is in a business organization, ask the employees. No one is closer to the source than people on the service and production lines—the billing clerk who observes paper waste in the invoicing process, the machine operator who sees metal waste in fabrication, the order-packer who follows company policy but knows the prescribed amount of packaging is excessive. Blue collar and white collar, employees are the true experts.

But if you want to prevent waste, start at the top. Effective waste prevention in a business organization begins with the commitment of top management to implement a program. Only management can empower everyone, all up and down the line. And only *middle* management can create the environment that enables employees at all levels to contribute to a shared objective to eliminate waste. In this chapter we examine how companies organize for such a mission.

## FIVE ORGANIZATIONAL STEPS TOWARD PREVENTING THE MAXIMUM AMOUNT OF WASTE

1. Top management announces the program and declares it a company-wide priority.
2. Employees form teams to plan, design, implement, and maintain the program.
3. Teams assess company operations for opportunities to prevent waste, set priorities and goals, and launch the program.
4. Results are monitored and fed back to all participants; the program is fine-tuned.
5. Accomplishments are celebrated and individuals recognized and rewarded for their good work.

### Support from the Top

Waste prevention in a business organization does not begin in earnest until it is recognized and declared valid by top management. But when that happens, the effect can be very powerful. In 1991, when 3M announced "Challenge '95," a 5-year program to improve productivity by minimizing waste, an internal memo from the

CEO left no room for doubt. "The effort must be cross-functional and be strongly supported by the General Managers and the entire Operating Committee," he underscored. "The stretch on the goals, particularly 10 percent unit cost reduction nondeflated, cannot be achieved by manufacturing without strong cross-functional support." The effort must be well organized and prioritized early, he said. It "must be in concert with other major corporate drivers" and "must have good measures in place." Challenge '95 placed primary emphasis on reducing toxic emissions to air and water. But it also called for reducing solid waste—35% worldwide by 1995 and 50% by 2000. The company was proceeding close to schedule by the end of 1996, surpassing 32%.

Waste prevention programs usually are part of a company's overall response to environmental protection. That's the case at Procter & Gamble. In a 1994 report, the chairman and CEO wrote: "P&G people around the world take their responsibility to the environment as seriously as their responsibility to shareholders. In fact, they're one and the same: we simply can't protect the interests of shareholders unless we do our part to protect the environment." The P&G Environmental Progress Report that year detailed, among much other data, steady reductions in the consumption of product packaging materials over the preceding four years—a 42% reduction in the use of paperboard packaging, 9% cut in corrugated cardboard, and a 14% reduction in plastics. It's important to note that these improvements in materials use occurred at the same time that overall production and sales were rising.

Coors Brewing Co. also demonstrates high-level support of waste prevention programs. Like many such companies, Coors has adopted a statement of principles governing its behavior in this area. Here is the opening statement of the company declaration adopted in 1993 by the board of directors:

> Coors Brewing Company believes that environmental, economic, and human needs can be met simultaneously. We believe that, individually and as a company, we can harness human creativity to continuously drive down pollution and waste, reducing both economic and environmental costs, and enhancing the health and vitality of our organization, employees, customers, and communities.

The statement then itemizes seven principles, including one that is directly relevant to our concerns: "We apply innovative technology toward the efficient use of resources, including reduce/reuse/recycle actions, waste minimization and the reduction of emissions to air, water and land." A case study later in this chapter illustrates a specific long-term initiative of Coors toward more efficient use of resources—the progressive reduction in thickness of aluminum beer can bodies.

Herman Miller, Inc., the office furniture company, was named one of the top ten environmentally managed companies by *Fortune* magazine in 1993. Here is the Herman Miller Environmental Awareness Statement:

As a corporate steward in our communities, through continuous improvement we will

- Minimize waste by following the priority order of reduce, reuse, recycle, compost, incinerate, landfill.

- Implement technologies to use energy resources efficiently.
- Strive to surpass conformance to the law, compliance being the minimum standard by which we rate our performance.
- Use company resources to promote environmental knowledge and awareness to those involved in our business, including employees, customers, regulators, suppliers, neighbors, and competitors.
- Review and improve the environmental impact of materials used in our products and processes.

Herman Miller set goals to be achieved by December 31, 1998, including eliminating the use of landfills; reducing material waste by 30%, "such waste defined as materials that enter the manufacturing facility but are not shipped to customers"; and reducing the volume of packaging materials for finished goods by one half.

## CASE STUDY: Coors Brewing Co.

Golden, Colorado

*Eliminating 12% of the aluminum in a 16-ounce beer can, Coors saves 673 tons of aluminum a year, material worth $1.5 million (1997).*

Coors Brewing Co. has been recognized for a number of years as a leader in waste prevention. In 1995 the company won the prestigious CONEG—Coalition of Northeastern Governors—Corporate Commitment Award for demonstrating leadership to integrate packaging waste reduction into sound management decisions and objectives. The recent further lightweighting of a 16-ounce beer can is one example.

Coors was the first U.S. brewer to market beer in aluminum cans. The first cans produced for market in 1965 weighed nearly twice what aluminum cans weigh today. Through a series of improvements, generally based on reducing the thickness of aluminum, an empty 16-ounce beverage can (including the top) now weighs about three quarters of an ounce. Minus the top, the average weight of 1,000 cans was 39 pounds in 1965 and 25 pounds in 1996. Forecast for 2000: 22.5 pounds.

An aluminum drink container is formed by drawing a round blank of metal first into the shape of a cup and then into the fully formed bottom and side-wall. The technical term is D&I—drawn and ironed. The final step is to form the neck, where the top lid, a separate piece, is attached. Reducing the amount of metal in the solid piece that forms the bottom and walls is a complex process. The objective is to reach the minimum thickness that will enable the container to retain its shape under pressure of the contents. Each time the can bottom is made thinner, for example, the bottom dome must be reconsidered and possibly redesigned to retain shape under pressure.

### Payback

Coors has been able to take lightweighting further than others in the industry. There is ample incentive to do so: each reduction of one-thousandth of an inch

**Figure 1**   LIGHT BREW, LIGHTER CANS. In 1965, 1,000 beer cans, minus the top, weighed an average 39 pounds; in 1996, 25 pounds. For Coors, each reduction of one-thousandth of an inch (0.001) in wall thickness of the 16-ounce can save $1 million in aluminum.

(0.001) reduces aluminum costs by $1 million. But the company believes it has gone about as far as possible in lightweighting the bottom and walls of the can. Further reductions in neck diameter may be possible but depend on acceptance by the buying public. Reducing the weight of a container yields a number of advantages aside from the initial saving in material costs. Transport cost also is reduced, for example; and chilling through a thinner can wall requires less energy.

## CASE STUDY: Herman Miller, Inc.

Zeeland, Michigan

*New wrapper for inter-factory shipment of chair components cuts packaging 70%.*

In 1990 Herman Miller began working with customers and suppliers on a source-reduction program, including the replacement of all single-use packaging with reusable packaging. Events at the Herman Miller Holland Chair Plant provide an example of how a returnable–reusable packaging system can work.

The company had been shipping shells for its Equa chair—about 300,000 a year—from Grand Rapids to Holland, Michigan, packed in lots of 28 inside a heavy corrugated box. Each shell was wrapped in a polyethylene (PE) plastic bag, and PE foam fillers were placed inside for added protection. PE was selected for both applications to facilitate recycling. After the one-way trip of 30 miles, all packaging was disposed by recycling.

The replacement system of logistical packaging does away with the corrugated carton entirely, as well as all expendable PE materials. Instead, chair shells are stacked on a plastic tray, capped with another plastic tray, and then covered with a corrugated sleeve. All these materials are returned to the source and reused. Implementing the new system took 6 months, including the appropriation of funds and manufacture and receipt of sufficient packaging inventory (dunnage) to fill the pipeline.

Herman Miller points out that adopting a system like this means adding many new "parts" to inventory. "The biggest challenge associated with returnable packaging has been keeping track of the dozens of unique parts," the company states. "It's a major problem if a supplier runs out of a returnable part, which in turn means delayed parts shipments, missed completions, and the use of more costly expendable packaging." *(A detailed discussion of returnable–reusable packaging may be found in Chapter 5.)* The company also notes that just-in-time sourcing "will remove any safety net, real or perceived, that may have been in place."

## Payback

Herman Miller requires any system change to pay for itself in less than a year. Adoption of the reusable dunnage described here cost about $500,000 and was recovered within the required time. In addition, the company lists these benefits:

1. Total cost savings, $1.4 million.
2. Packaging material reduction for this operation, 70%.
3. Damage claims reduced by 94%.
4. Handling ergonomics and safety both improved; 15–20 minutes handling time saved per unit of 28 chair shells.
5. Truck trailer cube improved by 15%.
6. Better use of warehouse space—no need for waste containers or storage of disposables.
7. Reusability: the corrugated sleeve can be used for 3 to 4 years and sustain 50 to 100 roundtrips.

3M, P&G, Coors, and Herman Miller all demonstrate the kind of commitment companies make when they strongly endorse waste prevention as a general objective. And while the primary audience is within the organization, a very important secondary audience are the many stakeholders outside, especially investors, consumers, and government regulators. Smart companies make certain their programs to prevent waste and the results of this good action are well publicized.

## Teams Empower Employees

The waste prevention team is the group of employees, authorized and empowered by top management, who plan, design, implement, and maintain the program. Obviously, the team is important to ensure that the many tasks associated with developing a program can be assigned and accomplished. But the team structure also is a very visible center of employee participation within the life of the company. Ideally, the team should become known as an easily accessible place for employees to contribute their time and talent.

Primary tasks of the waste prevention team are the following:

1. In cooperation with management, determining short-term and long-term priorities and goals.
2. Assessing company operations to identify specific opportunities for preventing waste.
3. Educating employees and inviting their participation in the program.
4. Designing and implementing specific waste prevention initiatives.
5. Monitoring progress, making adjustments, reporting to management.

How many members should the team have? Here's advice from the WasteWi$e program of the U.S. Environmental Protection Agency, in the publication *Business Guide for Reducing Solid Waste*:

> The size of your team should relate to the size of your company and be representative of as many departments or operations as possible. For a modest waste reduction program, an effective team might consist of just one or two people. The ideal candidate for a one-person team would be an individual who wears many hats and is familiar with the overall operations of your company. A two-person team might consist of a company manager and an administrative or technical support person. Larger businesses might opt to create a team of employees from different departments to encourage widespread input and support. These individuals can include environmental managers, building supervisors, technical or operational staff, administrative staff, maintenance staff, and purchasing staff, or other employees interested in waste reduction.

Herman Miller's Environmental Quality Action Team (EQAT) illustrates how a large company might organize. EQAT is a steering committee composed of 18 employees from throughout the company. It provides direction and oversight to all waste prevention activities. Five related groups help translate EQAT policy into action:

1. The Communications Committee carries the message to all company constituencies.
2. The Environmental Affairs Team manages day-to-day strategies. Among other members, this team includes a liaison person, whose job is to work closely with Herman Miller suppliers, and a specialist in minimizing solid waste.
3. The Green Building/Facility Design Group seeks the most sustainable practices and designs for company buildings.
4. The Design for Environment Team examines new products for environmental impact.
5. Packaging Engineering looks for less material-intensive ways to ship parts and furniture to Herman Miller customers.

## CASE STUDY: United Technologies Automotive

Berne, Indiana

*Interdepartmental teamwork at an auto industry supplier eliminates 180 tons of corrugated cartons, saves $1 million.*

The Berne plant of United Technologies Automotive (UTA) supplies exterior mirrors, grilles, and other painted products to the automobile industry. The plant employs 500 people. A recent cost-cutting program—designed and implemented by cross-functional teams—included special attention to reducing the cost of waste disposal related to the manner in which the factory received materials from suppliers. Everything arrived in periodic large shipments in corrugated containers, often 2 weeks of inventory requiring large amounts of storage space.

Plant management decided to ask suppliers to ship smaller quantities of components more frequently—just in time, or JIT—and to use a returnable–reusable container that would be designed and supplied by UTA. Participants in this change were the Purchasing Department, which developed the new container; the Materials Department, which changed delivery methods and truck schedules; and environmental managers.

UTA utilizes two types of returnable containers. One is a 6 × 6 × 20-inch polypropylene box, with tapered sides making it nestable. The other is a plastic tote made of high-density polyethylene (HDPE). The tote is used for shipping, storing, and moving work-in-process within the UTA plant. Totes nest onto a plastic pallet measuring 48 × 48 inches, slightly larger than a standard wooden pallet.

### Payback

By mid-1996, UTA had replaced 90% of incoming shipments with returnable–reusable containers; 60% of outgoing shipments were in returnables. No breakage had occurred. Containers have an estimated life of 10 or more years. Switching to plastic pallets yields savings of $400 to $500 a week, UTA says. The company estimates it recovered its investment in returnable–reusable containers in less than a year. Other benefits:

1. Greatly reduced overhead costs to purchase pallets and containers.
2. Reduced labor time to set up and break down corrugated cardboard containers.
3. A cleaner working environment.
4. Lower costs of inventory with the institution of just-in-time delivery—facilitated by returnable–reusable containers.
5. Easier cleaning of pallets with water or steam; wooden pallets virtually defied cleaning.
6. Lighter, easier to handle pallets: 45 pounds vs. 90 pounds for a wooden pallet of same size.
7. Reduced risk of injury to workers handling plastic pallets—no splinters, broken boards, or protruding nails. Reduction in back strains.

8. Rare breakage problems. When breakage occurs, plastic pallets, if reasonably clean, can be returned to the manufacturer for recycling and credit against the cost of replacements.

## Crucial Role of Middle Management

Frequently, the first member of a waste prevention team is a middle manager. Although most of the energy and creative thinking will come from nonsupervisory employees, the team's connection with overall company operations can be supplied most efficiently through a member of middle management. Such a person does not necessarily have to function as the team *leader*. His or her most important contribution to the team is the conferring of authority and the expression of company trust. Michael Sims, plant manager of Wainwright Industries, Inc., a recipient of the *Industry Week* "America's Best Plants" award, emphasizes this point. "Systematic" employee involvement "only occurs when the culture is based on sincere trust and belief in people," he is quoted in the magazine. "The most critical lesson learned ... is that the trust factor in organizations is controlled by middle managers, not senior leadership."

It's relevant to note that at America's "best" plants, as defined by *Industry Week*:

- 65% of the *total* workforce participate in self-directed teams.
- 90% of the *production* workforce are involved in empowered work teams.

## Priorities and Goals

The activity of preventing waste is likely to draw on many company resources: initially, employee time; later, perhaps some investment in capital equipment. And even though the central goal is very clear from the start—finding and eliminating waste—there probably will be peripheral goals, such as increasing operational efficiency or enhancing the company's reputation as a steward of natural resources and energy. Establishing priorities and setting goals will have to be determined cooperatively by management and the team. Since these priorities and goals will become part of the overall company operating plan, they will, like all other parts of the plan, be subject to periodic review and revision.

A sensible approach to establishing goals and setting priorities is to base them on a waste assessment (addressed in the next section). The New York City Department of Sanitation takes that approach in a published guide to businesses, and suggests considering the following when selecting targets and forming strategies for waste prevention.

1. Anticipated quantity of preventable waste: Begin with strategies that target those areas with the greatest potential for waste prevention.
2. Ease of implementation: Start with the easiest and most obvious steps. This will enable you to demonstrate immediate progress and provide momentum and justification for continued changes.

3. Initial cost and expected payback: Many waste prevention practices require an initial investment in equipment or personnel retraining. However, the investment often pays for itself in a short period of time—then the savings begin.
4. Employee cooperation: Consider the ease or difficulty of gaining cooperation for various waste prevention practices.
5. Relationships with suppliers: If you have a good relationship with suppliers, they may be more willing to meet your needs, especially if you are a large buyer.
6. Customer relations goals: Waste prevention practices that are evident to customers or the public will improve your company's image.
7. Flexibility in your waste disposal contract: Consider which practices will work best with your private carter.

## Assessment

Some years ago, Tom Peters introduced a new bit of shorthand to the business lexicon—MBWA, or "Management By Wandering Around." The idea is that effective management depends on knowing from personal observation what's happening in the company, walking around, viewing operations first-hand, and talking with employees to learn what's on their minds concerning production. The MBWA idea is simply a good business practice and known by many names in many variations. At EG&G Astrophysics, director of operations Jane Song instituted "routing by walking around," a process-improvement drill in which a person walks through a process as if he or she were a product being manufactured or a document being developed. At each stop along the way, the person asks, "Why am I stopping here?" and "How long will I stay here?" The same technique can be adapted to surveying the company landscape for opportunities to use less stuff for preventing solid waste everywhere in production and administration.

There are five ways to begin an assessment of the waste being produced by a business enterprise:

1. Examine records of the waste-hauling company.
2. Examine the company's purchasing records.
3. Walk through the facility.
4. Sort and analyze a targeted portion of the waste stream.
5. Sort and analyze the company's entire waste stream.

The chart "Alternatives for Waste Assessment" summarizes the advantages and disadvantages of each method listed above. No single method will yield all the information a company needs to pinpoint waste; neither will any single comprehensive analysis. Like a financial audit, a waste assessment is a snapshot at one point in time. It's the beginning of a continuous process.

A few observations about the methods:

- Examining records of the company that provides waste hauling services may provide useful information about gross tons of material taken to disposal. For example, if the company has a contract for removal and disposal of waste in

--------------------------------------------------------------------

## ALTERNATIVES FOR WASTE ASSESSMENT

| Method | Advantages | Disadvantages |
|---|---|---|
| Examine Hauler Records | • May provide accurate data on weight or volume of waste.<br><br>• Can require less time and effort than a walk-through or waste sort. | • May not provide adequate data if accurate records do not exist.<br><br>• Not likely to provide information about specific waste components. |
| Examine Purchasing Records | • Can track potential waste from the source.<br><br>• Can require less time and effort than walk-through or waste sort.<br><br>• Can be most accurate for tracking small items, low-volume and occasional materials. | • Not likely to provide the full picture.<br><br>• If purchasing is not centralized, may be incomplete or difficult to collect. |
| Walk Through the Facility | • Can require less time and effort than a waste sort.<br><br>• Allows a close look at operations.<br><br>• Can provide qualitative data about waste generation.<br><br>• Allows interviews with workplace personnel, eliciting insights not found in records or waste sorts. | • May not identify all wastes.<br><br>• May not be representative if conducted only once.<br><br>• Does not provide precise data about quantities of wastes. |
| Sort Targeted Waste | • Produces quantitative data about specific types of waste or functional areas.<br><br>• Takes less time and effort than comprehensive waste sort. | • May omit major components of facility's waste.<br><br>• May not be representative if conducted only once. |
| Sort All Waste | • Provides waste generation estimates for entire facility.<br><br>• Provides quantitative information for each waste component. | • Requires significant time and effort.<br><br>• May not be representative if conducted only once.<br><br>• Does not provide qualitative information on why wastes are generated. |

Adapted from U.S. EPA WasteWISe

--------------------------------------------------------------------

**Figure 2**   BENCHMARKING. Records and research can be used to assess a company's level of waste production at any time during a waste prevention campaign.

containers dedicated solely to the company's account (such as a 20-, 30-, or 40-yard "roll-off" box), the hauler should know exact weights, since most disposal sites charge by the ton. But if waste collected at one company is mixed with waste from other sources, it may be impossible to determine what came from where.

- Records of the company purchasing department should yield reliable data about a significant portion of material entering the plant—quantities of copier paper received, or tons of steel coil, for example. Knowing how much material has arrived at the company adds perspective to how much of the same kind of material leaves as waste. But many materials delivered to the company do not pass through the purchasing department. Pallets, corrugated cardboard boxes, and other logistical shipping materials, for example, may arrive in great quantity and become a significant disposal expense.

- A facility walk-through can provide a good initial estimate of the potential for waste prevention. This technique also can greatly improve information from other sources, such as data from the waste hauler: a quick look in the dumpster ("Management by Lifting the Lid") will show what *kinds* of materials are being sent to disposal. A walk-through may enhance purchasing statistics by spotlighting unnecessarily large inventories of certain materials, such as corrugated boxes: large supplies of a large variety of box sizes may indicate an opportunity to adopt a smaller number of standard sizes. The downside of a walk-through is treating it as a one-time check-off on the way to waste prevention. MBWA should become a habit.

- Waste sorting, either in a targeted area or for the whole facility, yields the most reliable information about what's being discarded at a given point in time; and as any experienced dumpster diver will report, a waste audit produces a vivid, detailed picture. Sometimes, data on what's in the trash can be fed back to organizational departments with benefit. At Walt Disney World, if an extraordinary number of promotional brochures land in the trash, the department that produced them is notified, along with the suggestion that printing quantities may need to be adjusted. But unless the waste stream is closely monitored on a continuous basis (it is at Disney), waste sorting, which is an expensive technique, yields perishable data, just another snapshot.

Possibly the most effective way to gather useful information is to combine a walk-through with one other method of assessment, such as an examination of purchasing records. To complement hard numbers from purchasing, it will be important to document the walk-through, recording the team's observations and ideas on paper. Price Waterhouse documents its JIT walk-throughs with a video camera, taping examples of inefficiency and waste as they go. "This is a simple and persuasive way to record problems or opportunities," Alexander Hiam writes in *The Vest-Pocket CEO*, adding that video is especially recommended if proposed changes must be presented to people unfamiliar with the facility.

## Educating Employees, Encouraging Participation

Kaoru Ishikawa, the father of Japanese quality control, once said, "In QC, one cannot simply present a goal and shout 'work hard, work hard.'" The same can be said of preventing waste at the source. It is not enough to say, "Let's do it!" Recognizing and rooting out waste is a learned skill, in part because we're looking at familiar territory, the place where we spend the majority of our waking hours. It may be quite a comfortable place; oftentimes we know it so well we could find our way around in the dark. And that, of course, is the essence of the problem. We must turn the lights on bright and really look at this place, as if for the first time.

First official word about the company's waste prevention program should come from the president or top resident manager. If possible, an oral announcement is best. This will give employees a chance to ask questions—What does waste prevention mean? How do employees fit into the picture? Why is the company doing this now? What's in it for us and for the company? Who's going to be in charge? What can we do to help? A complete announcement will answer all those questions. With or without a live announcement by the CEO, coverage in the internal newsletter, bulletin board postings, and quality circles is essential to let everyone know what's happening and how they can play a role.

Communication and education must be viewed as continuous processes. At Grote Industries, Inc., in Madison, Indiana, communication begins afresh each week with the 1 p.m. Tuesday meeting of the waste prevention team, rarely a long meeting but long enough to cover what's new and give department representatives a chance to talk with one another about waste problems and toss around possible solutions. Rick Wardrip, environmental engineer at Grote, recalls how one little instance of waste got remedied in the weekly gathering. A team member from the maintenance department reported with some frustration that 5-gallon buckets of machinery grease were being tossed in the trash half full. How come? A little investigation revealed that the pump used to remove grease from the bucket only reached half way down. There appeared to be no ready solution to that problem. But Grote team members also knew that the painting department used grease—new grease—as a sort of adhesive to hold disposable plastic sheeting against walls of the paint booth. Putting one department and another department together, the team suggested that half-empty buckets from maintenance be routed to painting, where the remaining half-bucket of grease could be scooped out and put to good use.

Over a few years, a steady stream of little improvements like the Grote grease bucket can turn into significant savings. Within just one year at Grote, for example, waste prevention reduced the landfill disposal bill 60%, from $45,000 to $18,000.

## Suggestion System: Priming the Pump

Many companies either currently use or have a history of using a formal suggestion system to encourage employees to contribute ideas for improving operations. Indeed, the old suggestion box has been around a long, long time! Many companies offer monetary rewards to employees for their suggestions; some do not. Many offer rewards only for implemented suggestions; some reward all suggestions. There are countless variations. We'll examine methods to help people start focusing on opportunities for waste prevention, how management can assist employees in perfecting the presentation of their ideas, and illustrate how specific companies manage their suggestion systems. Much of the discussion will be based on the Japanese philosophy of continuous improvement, including some illustrations from the Toyota Motor Manufacturing plant in Georgetown, Kentucky.

One of the most comprehensive discussions of suggestion systems is in *The Idea Book: Improvement Through TEI (Total Employee Involvement),* edited by the Japan Human Relations Association (Productivity Press, Cambridge, 1988). About launching a suggestion program, the book offers this counsel:

Start by looking for examples of waste, inconsistency, or inadequacy that are sure to be around you. The important thing is to identify problems close to you and solve these routine problems one by one. Even people who have won company or national-level awards started out by simply eliminating the waste, inefficiency, or inconsistency they saw around them. Steady effort in these areas lays the groundwork necessary for coming up with major improvement suggestions.

There is a wealth of good advice in this short passage. Notice the emphasis on paying attention to your immediate surroundings—the place you should know best—and on looking closely at routine things, eliminating waste one step at a time. Start small, make incremental improvements. Don't expect to "hit one out of the park" your first time at bat. Be patient. With experience you will begin to see opportunities for substantial improvements.

## Thought Starters

Suppose an employee is enthusiastic about waste prevention, wants to contribute suggestions, but just doesn't know how to start. "I've thought about it, but I don't seem to have any ideas," he or she says. *The Idea Book* presents a system for formulating ideas called "The 5W2H Method," a simple and easy-to-remember drill. The five W's stand for What, Why, Where, When, and Who; the two H's are How and How much. The idea is to question everything, from every possible angle.

One of the best questions to keep in mind about anything observed in the company is, "What would happen if ...?" For example, What would happen if ...

... We took the outer wrapper off the product?
... We told suppliers we want them to stop using stretch wrap on incoming freight?
... We reduced the trim edge 50%?
... We enlarged (reduced) the package 100%?

In any manufacturing setting there must be hundreds of such "What if" questions to ask. If employees are encouraged to ask bold questions like these from the very start, and given a system for developing answers, they will make a good beginning on the quest toward waste prevention. The table headed "How To Formulate Ideas, the 5W2H Method" shows how question-asking can be directly related to the activities that occur in a manufacturing setting.

## Launching the Waste Prevention Program

With a team in place, the mission defined, and priorities clear, what happens next? If an initial waste assessment also has been completed, the team may already have identified some targets for immediate action. It may now be clear, for example, that the office copying machines should be reprogrammed to automatically copy on both sides of the paper unless the default setting is overridden. Maybe it has become obvious that suppliers should take back pallets and other logistical shipping materials—or find a way to eliminate them. Implementing such examples, as well as most other waste prevention initiatives, will require at least employee time and

------------------------------------------------------------

## HOW TO FORMULATE IDEAS, THE 5W2H METHOD

| Question | Focus | Statement | Counterstatement |
|---|---|---|---|
| **WHAT?** | General subject | What is being done? Can it be eliminated? | Eliminate unnecessary tasks, materials |
| **WHY?** | Purpose | Why is this step necessary? Clarify the purpose. | |
| **WHERE?** | Location | Where is it being done? Does it have to be done there? | |
| **WHEN?** | Sequence | When is the best time to do it? Does it have to be done then? | Change the sequence or combination |
| **WHO?** | People | Who is doing it? Should someone else do it? Why am I doing it? | |
| **HOW?** | Method | How is it being done? Is this the best method? Is there some other way? | Simplify the task |
| **HOW MUCH?** | Cost | How much does it cost now? What will the cost be after improvement? | Select a better, less costly method |

Adapted from The Idea Book.

------------------------------------------------------------

**Figure 3**  THOUGHT STARTERS. When employees are stuck for waste prevention ideas, asking the five Ws and two Hs can reveal possibilities. From *The Idea Book: Improvement Through TEI (Total Employee Involvement)* edited by the Japan Human Relations Association. English translation copyright © 1988 by Productivity Press, PO Box 13390, Portland, OR, 97213-0390, (800) 394-6868. Reprinted with permission.

perhaps also money and space. When and how should the cost of action be compared to the possible payback? Figures 4 and 5 address this question.

The forms titled "Evaluating the Economics of a Waste Prevention Option" and "Evaluating the Technical Merit of a Waste Prevention Option" provide a framework for thoroughly examining a waste prevention option. On the technical side, the first question is the most important: "Has this option been proven in service?" In other words, has the option already been tested within the company, even on a limited basis? This implies that waste prevention ideas will be tested, at least to some degree, before they are formally proposed for general adoption. There are several good reasons for taking this practical approach rather than reviewing untested suggestions for improvement.

## EVALUATING THE ECONOMICS
## OF A WASTE PREVENTION OPTION

Company _____

Date _____

Evaluator _____

> INSTRUCTIONS: Check the appropriate response for each question.
> Skip questions that do not apply.

WASTE   REDUCTION   OPTION _____

|  | YES | NO | NOT SURE |
|---|---|---|---|
| 1. Is this option within price range considering both capital and operating costs? | | | |
| 2. Does this option have an acceptable payback period? | | | |
| 3. Does this option reduce raw materials cost? | | | |
| 4. Does this option reduce utilities cost? | | | |
| 5. Does this option reduce storage and handling costs? | | | |
| 6. Does this option reduce regulatory compliance costs? | | | |
| 7. Will this option reduce the costs associated with worker injury and illness? | | | |
| 8. Will this option reduce insurance premiums? | | | |
| 9. Will this option reduce waste disposal costs? | | | |

Source: Chamber of Commerce, St. Joseph County, Indiana

_____

**Figure 4**   EVALUATION, STEP 1. What are the economic merits of a waste prevention initiative?
The nine questions posed here will provide a fairly complete answer.

## EVALUATING THE TECHNICAL MERIT
## OF A WASTE PREVENTION OPTION

Company _____

Date _____

Evaluator _____

> INSTRUCTIONS: Check the appropriate response for each question.
> Skip questions that do not apply.

WASTE    REDUCTION    OPTION _____


                                                    YES      NO      NOT SURE


1. Has this option has been proven in service?

2. Will this option maintain product quality?

3. Will this option adversely affect productivity?

4. Are you certain that this option will produce less waste?

5. Will this option require additional staff?

6. Is plant layout capable of incorporating this option?

7. Will this option improve worker safety?

8. Are materials and parts readily available?

9. Can this option be easily serviced?

10. Are other businesses using this option?


Source: Chamber of Commerce, St. Joseph County, Indiana

**Figure 5**    EVALUATION, STEP 2. What are the technical merits of a waste prevention initiative?
Like questions about economic merits, technical merits may be impossible to assess
until the idea has been tried.

1. The full cost of implementing an idea—labor, materials, space, etc.—will be much more apparent after a pilot project.
2. Hidden flaws with a proposal are likely to surface when it is put to a test.
3. If rewards are given for ideas, requiring partial implementation adds discipline to the process and keeps the suggestion–review queue from getting too long.

The waste prevention team plays a very important role in nurturing the idea-development process. Team members create the climate in which all employees feel comfortable about participating. (At Toyota Motor Manufacturing Kentucky, Inc., one team member makes a point of periodically visiting workers on the third shift, 10:30 P.M. to 7 A.M., to encourage these people who are somewhat removed from the daytime-mainstream to participate. "If you have ideas, page me and I'll call you back," he says.) Team members can serve as mentors and coaches to other employees who are in the process of implementing ideas. An orderly process will have a number of steps, including the initial approval to proceed from a supervisor; the assembly of materials and equipment, if necessary; communication with vendors and customers, if necessary; the designation of a trial period and careful monitoring of results; and finally, analysis and preparation of a written report.

## Putting Suggestions in Writing

Here is where the momentum of a new waste prevention program may encounter its first slow-down. An implemented idea for preventing waste may prove to be very workable and produce significant savings, but if it cannot be translated into a complete and convincing written statement (with illustrations, where useful, to enhance the words), it may be essentially worthless. A clear, complete, but concise statement of the suggestion is important for two reasons:

1. To the suggester, a complete written statement may mean the difference between receiving a monetary award for a valuable improvement or having it rejected out of hand. Managers and others reviewing the suggestion may be familiar with how it evolved. But even if they do know the background, the only fair way to compare one suggestion to others is to require that all be evaluated on the same basis, as write-ups.
2. To the company, a good write-up facilitates passing good, implemented ideas along to employees and departments throughout the organization.

Most suggestion systems prescribe a form for the write-up (see the Toyota example). But a form does not necessarily make the task any easier for employees who do not routinely write as part of their work. Here is another place where the waste prevention team can perform a very important service—as coaches to employees when it comes time to do the write-up. When a suggester is having difficulty, often he or she will say, "I know how it works but I can't seem to get it onto paper." One technique a coach can use is to sit down with the suggester—and a tape recorder—and with a series of sequence questions (What happens first? What next? What next?) lead the employee through the entire process that is being improved, both the before and after situations. Later, when the employee is writing the sug-

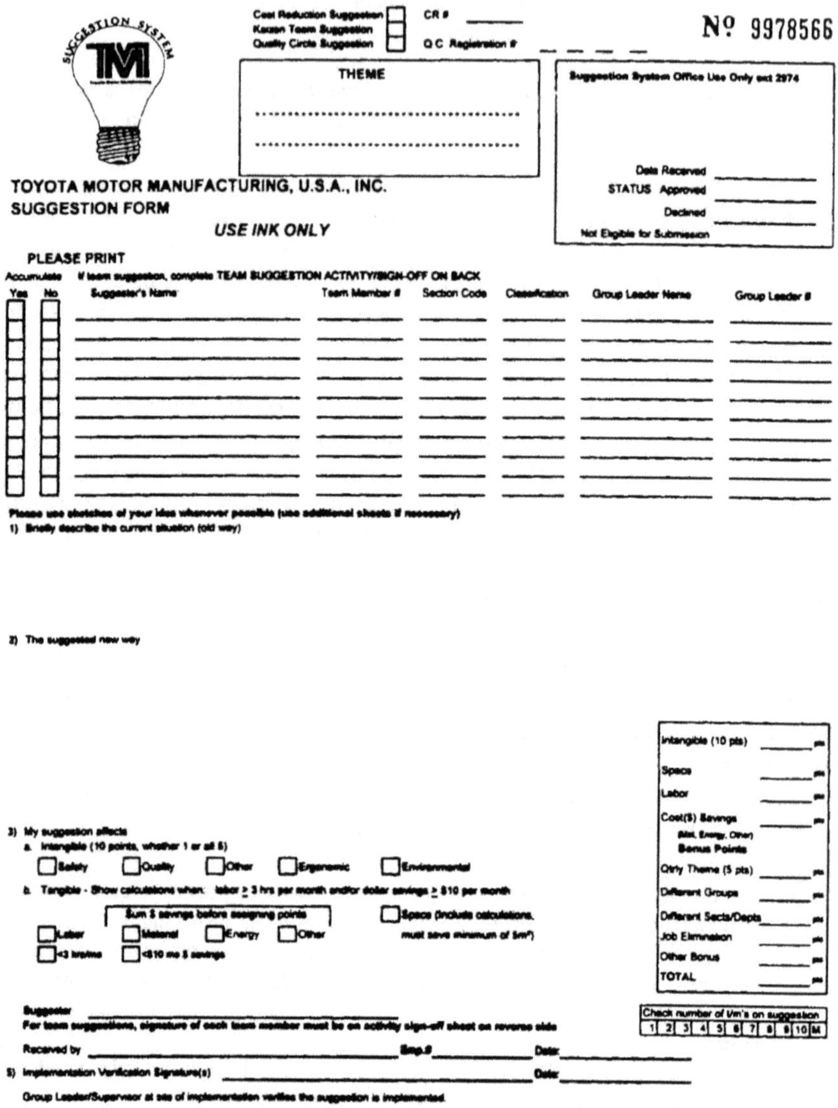
SUGGESTION SYSTEM TMM

Cost Reduction Suggestion ☐ CR # ______
Kaizen Team Suggestion ☐
Quality Circle Suggestion ☐ Q C Registration # ____

Nº 9978566

THEME
..........................................
..........................................

Suggestion System Office Use Only ext 2974

Date Received ______
STATUS Approved ______
Declined ______
Not Eligible for Submission

TOYOTA MOTOR MANUFACTURING, U.S.A., INC.
SUGGESTION FORM

*USE INK ONLY*

PLEASE PRINT

Accumulate — If team suggestion, complete TEAM SUGGESTION ACTIVITY/SIGN-OFF ON BACK

| Yes | No | Suggester's Name | Team Member # | Section Code | Classification | Group Leader Name | Group Leader # |
|---|---|---|---|---|---|---|---|
| ☐ | ☐ | | | | | | |
| ☐ | ☐ | | | | | | |
| ☐ | ☐ | | | | | | |
| ☐ | ☐ | | | | | | |
| ☐ | ☐ | | | | | | |
| ☐ | ☐ | | | | | | |
| ☐ | ☐ | | | | | | |
| ☐ | ☐ | | | | | | |
| ☐ | ☐ | | | | | | |
| ☐ | ☐ | | | | | | |

Please use sketches of your idea whenever possible (use additional sheets if necessary)

1) Briefly describe the current situation (old way)

2) The suggested new way

3) My suggestion affects
a. Intangible (10 points, whether 1 or all 5)
☐ Safety ☐ Quality ☐ Other ☐ Ergonomic ☐ Environmental
b. Tangible - Show calculations when: labor ≥ 3 hrs per month and/or dollar savings ≥ $10 per month
Sum $ savings before assigning points
☐ Labor ☐ Material ☐ Energy ☐ Other ☐ Space (include calculations, must save minimum of 5m²)
☐ <3 hrs/mo ☐ <$10 mo $ savings

| | |
|---|---|
| Intangible (10 pts) | ___ pts |
| Space | ___ pts |
| Labor | ___ pts |
| Cost($) Savings (Mat, Energy, Other) | ___ pts |
| Bonus Points | |
| Qtrly Theme (5 pts) | ___ pts |
| Different Groups | ___ pts |
| Different Sects/Depts | ___ pts |
| Job Elimination | ___ pts |
| Other Bonus | ___ pts |
| TOTAL | ___ pts |

Check number of tm's on suggestion
1 2 3 4 5 6 7 8 9 10 M

Suggester ______
For team suggestions, signature of each team member must be on activity sign-off sheet on reverse side

Received by ______ Emp.# ______ Date ______

5) Implementation Verification Signature(s) ______ Date ______

Group Leader/Supervisor at site of implementation verifies the suggestion is implemented.

**Figure 6** INTIMIDATING? Suggestion forms may be a hurdle for some employees. Coaching helps.

gestion, he can replay the tape as a reminder, in his or her own words, of what the improvement is all about. The addition of comparative, before-and-after data about cost and waste will round out a complete suggestion.

## Keeping a Waste-Prevention Journal

Ideas about waste prevention are likely to dawn at any time. Doing something about them may not be opportune until later, however, and by then some of the

---

IMPROVING THE WRITE-UP
OF AN IMPLEMENTED SUGGESTION

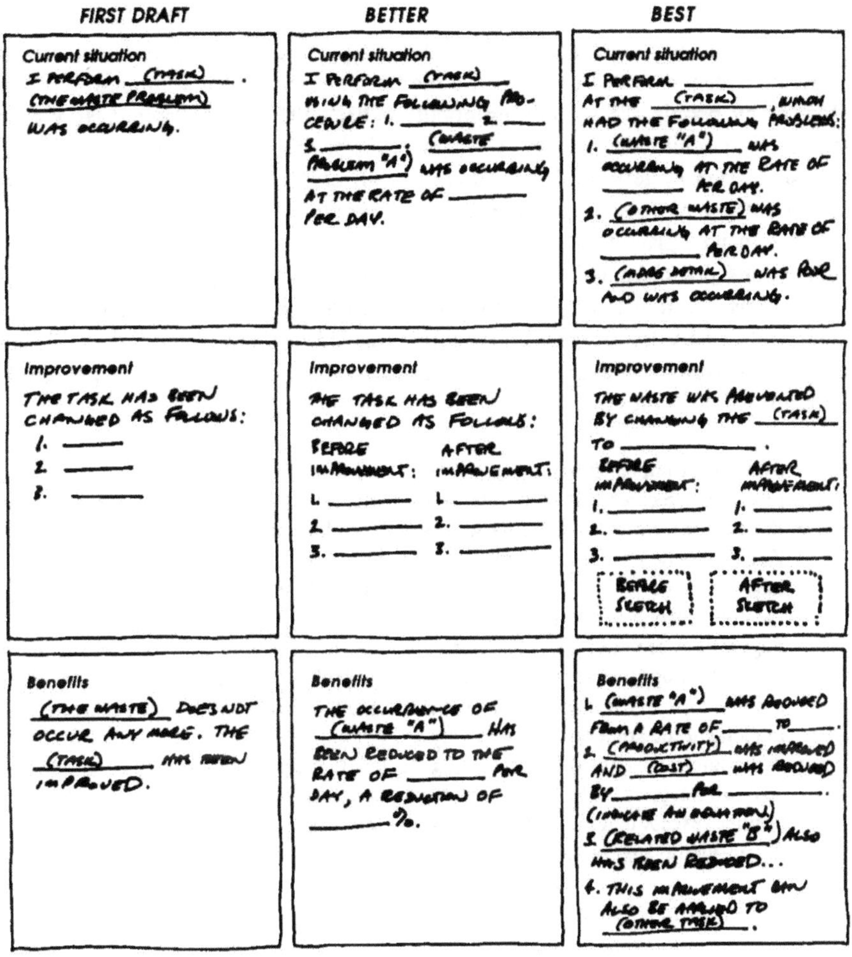

Based on an illustration in The Idea Book

**Figure 7** Adapted from *The Idea Book: Improvement Through TEI (Total Employee Involvement)*, edited by the Japan Human Relations Association. English translation copyright © 1988 by Productivity Press, P.O. Box 13390, Portland, OR 97213-0390, (800) 394-6868. Reprinted by permission.

details may have slipped from memory. To help employees capture ideas as they occur, some companies suggest they keep a journal or notebook devoted exclusively to waste. *The Idea Book* suggests keeping a notebook, and offers this practical advice: "Don't kill an idea at the beginning, saying that something won't work. Write it down first. You never know what will come in handy later. But once you write down a fact, delete it from your consciousness so that you can devote your awareness to making new observations."

## Setting Targets for Idea-Production

How many waste prevention ideas should an employee, a shift, or a department generate over the course of a week, a month, or a year? That's impossible to say. How many people are we talking about? What's the starting point? How much support will employees receive? Each case will be unique. At Toyota, for example, the suggestion system guide distributed to all employees encourages "the habit of submitting at least one suggestion per month. Often working through one suggestion will provide insight for yet another idea. The long-term goal is two suggestions per team member per month." Of course, Toyota is talking about suggestions for improving all aspects of a complex automobile manufacturing plant, not simply the prevention of solid waste. But many details of the Toyota suggestion system could be adopted as-is by any company waste prevention program, such as these:

- Time spent implementing a suggestion is paid time. But background work such as researching records, gathering data, and developing and writing ideas is not paid time.
- Suggestions are one of two basic types, depending on the nature of the benefit. *Tangible* benefits save materials, energy, company labor, and contracted outside labor. *Intangible* benefits relate to safety, quality, the environment, ergonomics, or the conservation of work space.
- Suggestions are paid after implementation. Merely pointing out a problem is not sufficient. Suggesters must think through the problem to the point of developing a solution, even if they do not have complete information or the expertise to implement the solution independently.
- Not eligible for consideration are suggestions that offer no constructive solution, deal with supplier selection or price, or concern policies, regulations, standards, or rules (except that a suggestion for waste prevention might demonstrate the need to consider changing such matters).

## Waste Prevention Blitz

Some companies have found it beneficial to hold an annual "blitz"—a concentrated period such as 20 consecutive days when all employees are encouraged to contribute simple, basic ideas about waste prevention at the rate of one a day. At a Toyota subsidiary in Japan, for example, during the 20-day blitz every worker carries a pocket notepad with a cover imprinted, "One Improvement a Day." When an improvement is made, he or she writes it up on a page and tears that page off. Those pages are displayed on a bulletin board as credit to the employee and incentive to others. Although some actions may appear more like routine housekeeping chores than genuine process improvements, all are recognized during the blitz.

## Monitoring, Reporting, and Rewarding Waste Prevention

The waste prevention team at Grote Industries wanted to show its appreciation to the entire production line for excellent cooperation. So, with only a little advance work and at minimal cost, the team arranged to deliver a cold soft drink to everyone on the line, all shifts, on a hot summer day. The can was nestled in an insulated holder imprinted with the company logo and a few words of thanks from the team. Sometimes it's the thought that counts most, and making a little reward a surprise has a way of magnifying good feeling.

Monitoring and reporting the successful waste prevention accomplishments of employees can have much the same effect. This is a dynamic process in which success builds upon success. It is crucial for the team to keep a central record of accomplishments as individual suggestions for waste prevention are implemented. Statistics are vital—pounds or tons of material eliminated, dollars of outside cost saved, hours of labor freed up for more productive use, ergonomic improvements, etc. There should be enough data available after a year or so of experience with a waste prevention program to make a persuasive statement about what has been accomplished. If employees are being rewarded with cash for suggestions, general in-plant publicity (newsletter, bulletin boards) is highly appropriate.

### Monetary Awards

If money is being awarded for suggestions, how much is appropriate? The system at Toyota's Kentucky plant provides an example. Toyota uses a scale to relate dollar savings per month to the award of "points." Up to 499 points, the award is a Sears gift certificate. For 500 points and higher, the award is cash. Here is the table that applies to tangible suggestions for preventing the waste of materials, energy, and other nonlabor costs:

| Savings Per Month | Award Points |
|---|---|
| <$10 | 3 |
| $10–29 | 5 |
| $30–59 | 10 |
| $60–99 | 15 |
| $100–149 | 25 |
| $150–199 | 35 |
| $200–249 | 50 |
| $250–299 | 70 |
| $300–349 | 100 |
| $350–399 | 150 |
| $400–699 | 250 |
| $700–999 | 400 |
| $1,000–1,299 | 550 |
| $1,300–1,699 | 750 |
| $1,700–2,999 | 1,100 |
| $3,000–4,499 | 1,500 |
| $4,500–5,999 | 2,000 |
| $6,000–7,999 | 2,500 |
| $8,000–9,999 | 3,000 |
| $10,000+ | 3,000 |

Each additional $1,000 saving per month yields 30 additional points. The maximum payment is $10,000. Other tables apply to labor savings, intangible suggestions, and group bonuses—it's an intricate system! But it seems to stimulate participation. In 1996, 75,040 suggestions were developed by the 10,000 workers at the Toyota plant in Kentucky. Awards totalled $2,331,500, and the company assumed whatever local, state, and federal tax liability applied. While Toyota prizes individual initiative in waste prevention, team suggestions are valued highly and rewarded well: each team member receives 50% of what an individual would have received, up to 10 members. The administrator of the suggestion system makes an important point: no one is ever laid off as a result of a suggestion. Thus, if a labor-saving suggestion eliminates five jobs on a shift, as it has happened at Toyota, the employees are moved to other jobs.

The philosophical underpinning of the Toyota suggestion system can serve as a model for all companies. "The best improvement ideas come from the people who do the job every day, the hands-on people, not the managers and engineers," the company guide states. "People are most likely to use their creativity when their ideas are put into practice and when they are given recognition for their ideas."

### CASE STUDY: Toyota Motor Manufacturing Kentucky, Inc.

Georgetown, Kentucky

*Suggestion system yields a housekeeping idea that saves $6,000-plus a week.*

Toyota uses a lot of rags and mops to keep production areas clean. Disposing of dirty rags and mop-heads can be costly if they contain oily material, which virtually all do because of the prevalence of oil and grease in an automobile assembly plant. For example, to dispose of one gaylord box full of dirty rags and mop-heads as hazardous waste will cost from $300 to $500. (A gaylord box is a heavy-duty corrugated cardboard bulk container with a volume of approximately 1 cubic yard.)

But here's a twist: If these same dirty textiles are sent to an industrial laundry for washing and reuse, under federal regulations they are not considered hazardous waste. This difference was apparent to members of the Toyota environmental management team, who suggested trying the alternative method. The demonstration worked fine. Experience shows the laundered rags and mopheads can be reused up to 10 times before they must be disposed of as hazardous items.

Previously, Toyota was generating between 20 and 30 gaylord boxes of contaminated cleaning materials each week. At $300 per box, the company was paying $6,000 to $9,000 a week for disposal. That cost has been reduced to a small fraction by laundering and reusing the rags and mop-heads. (A Toyota manager said the company also made certain that the outside laundry service was not discharging contaminated washwater to the public wastewater treatment system.)

# CHAPTER 3

# The Front Office—
# A Highly Visible and Easy Target

With the CEO on board and a mission statement adopted, the company waste prevention campaign is set to begin. And the first target of opportunity is just down the hall from the corner office at the copying machine. Not long ago, the hum of the copy center was the *sound* of business, and the widespread distribution of thick paper reports was the *look* of business. But that was yesterday. Today, we're smarter. Before deciding how many copies to print, waste conscious managers ask, Who really needs this report in this form? Is there a better way to distribute it—on the Internet, perhaps? Is there any good reason *not* to double-side it? Why not print just a few copies and route them? Remarkable savings begin to occur when people start asking such questions.

- At Bell Atlantic, the campaign to reduce copier paper usage achieved in 1 year a reduction of 400,000 pounds of paper, saving $300,000.
- Chrysler eliminated 287,000 pounds of copier paper over 2 years. That's about 28 million sheets.
- BankAmerica cut the production of detail reports by 90%, saving $185,000 in staff time that would have been required simply to review those reports.
- At Home Box Office (HBO) headquarters in New York City, 30 cubic yards of paper were eliminated per year by strict application of the duplexing rule: all documents are copied on both sides of the page unless single-sided copies are specifically requested—and justified.
- The Walt Disney Company figures that if half of all its copier production is double sided, the company saves $300,000 a year.

How much paper are we talking about? The U.S. EPA reports that the average office worker generates 52.3 pounds of office paper per year. Other data suggest that is a conservative estimate. Nonetheless, it's a lot—more than 10 reams, or a stack of sheets about 20 inches tall. There are several major kinds of preventable waste in offices, but none is easier to identify or faster to yield than paper, especially photocopy paper, and we open this chapter with a close look at effective strategies, including e-mail. Then we'll turn to waste prevention opportunities in office supplies, the purchasing function, food service, the mail room, and housekeeping.

## STRATEGIES FOR THE COPY CENTER

Books, newspapers, magazines, and tax forms are printed on two sides of the sheet of paper. Why isn't photocopying routinely handled that way? The answer, of course, has to do with limitations of machinery, paper, and time; the notion that single-sided is somehow more elegant, hence more appropriate, for some kinds of copying; and professional barriers—perceived, at least—to double-siding things like contracts and other legal documents. Let's take a look at these concerns.

### Copying Machines

The office machine industry classifies photocopiers according to the number of copies they will print per minute, and speed generally correlates to duplex copying ability. The smallest, least expensive machines produce less than 10 copies per minute and are not built to print automatically on both sides of the sheet, although an operator may be able to achieve duplexing manually. The largest, most expensive machines run at 90 or more copies per minute and will automatically duplex on demand. In general, personal and convenience copiers—machines with speeds under 45 copies per minute—account for less than 20% of duplex copying. Most duplexing is done on the larger, faster equipment in staffed copy centers.

In a report on reducing office paper waste, Inform, Inc. concludes, "The largest increases in number of pages duplexed could be achieved by focusing efforts on high-capacity machines, typically found in centralized copying facilities." Though they constitute only a small proportion of all photocopiers, the big, high-speed machines consume the majority of paper. And they are much more reliable in the duplexing mode than they were a few years ago, due in part to improved designs for handling the inherently complicated process of printing on two sides of a sheet, and better understanding by machine operators of the need for care in storing photocopy paper and preparing and loading original copies. (The Inform report also points out that single-spacing a document rather than double-spacing has the same impact on paper use as the decision to print two-sided instead of one-sided.)

### Programming Copiers to Duplex Automatically

Companies have programmed their large photocopy machines to automatically print on two sides unless this default setting is overridden by the operator. Results have been mixed. If the operator knows the machine is set up to duplex and can easily select one-sided as necessary, programming machines this way should work without difficulty. Awareness is key. Several years ago at AT&T's Bell Labs, management decided to try making double-sided copying the automatic setting for its self-service copiers. But results were poor. "If users didn't consciously choose single-sided, they got double-sided, which they often didn't want," Fran Berman writes in

*Trash to Cash: How Businesses Can Save Money and Increase Profits.* "Waste paper volume actually increased until finally they went back to the old standard." Returning to this initiative a second time, the company public relations department pitched in with a poster campaign encouraging double-sided copying, with new posters every 3 months. A continuing employee education campaign plus the advent of more user-friendly photocopying equipment have helped Bell Labs to establish duplex copying as the company standard.

## Charging More for Single-Sided

If copying is charged back to departments, establishing a lower rate for double-sided may improve duplexing volume. The differential might be half a cent, for example, with single-sided charged at 3 cents a copy, double-sided at 2-1/2 cents. Columbia University, for example, charges more for single-sided copying (see case study).

## Replacing Paper Copies with e-Mail

For years, companies with far-flung operations thought there was no escaping the high cost of communicating on paper and paying the freight, usually by air, for distribution. The Internet has revolutionized communications among company locations. Consider the Lockheed Martin Corp. Headquartered in Bethesda, Maryland, the company uses the World Wide Web to distribute policy and operations data to 80,000 employees nationwide. Employees use their desktop computers to quickly search and download company information. The risk of unauthorized outside access is reduced by the use of passwords.

"The Internet has proven to be a cost-effective and timely way to distribute corporate policies and procedures," Nancy Corder, director of corporate policies and procedures states in *WasteWi$e Update* of the U.S. EPA. Lockheed Martin reduces costs a number of ways, Corder says. It takes less time for employees to search for information with a computer than by paper-based methods. But the most impressive savings are in paper and postage. Corder calculates the cost to print and mail one 100-page policy manual to 80,000 employees is $250,000. Internet distribution also makes revisions and amendments much simpler. Making changes to a staff benefits booklet—an annual rite at many companies—takes much less time and money. In addition, the electronic network can be used to alert employees when revisions occur.

## The Fax Machine Factor

Fax machines that print on thermal paper contribute to paper waste because thermal paper fades over time, and fax recipients who want to be assured of having a permanent copy will usually make a photocopy of a thermal-paper fax. There are two ways to address this problem. One is to replace the thermal-paper fax machine with a machine that prints on plain paper. This may seem an extravagance until you compare prices between new machines and older models. If an existing, thermal-paper fax machine is more than 5 to 7 years old, a new machine that prints on plain paper may cost only half as much. The other solution is to take the fax machine out of the loop by incorporating fax receipt and transmission into desk-top computers, avoiding the use of paper altogether. But even where that capability exists, it's unlikely the fax machine can be removed entirely.

The office fax machine also falls prey to unwanted messages—"junk faxes"—from advertisers. Companies use a variety of methods to minimize the volume of unwanted messages, including turning the machine off after hours, changing the fax number frequently, adopting an unpublished fax number, and even assigning a 900 number to the machine, causing the fax sender to pay for the call!

### The One-Page Rule, and Other Good Ideas

One of the best overall ways to reduce paper consumption is to be brief. No letter or memo should be any longer than essential for effectiveness—no more than one page, if possible; no form should demand any more detail than necessary. The advice seems obvious but is often ignored. Somehow, length is equated with substance. Frequently the opposite is true—undue length reveals poor preparation for writing and little attention to the organized presentation of ideas. But there is another, much more compelling reason to keep written communications concise. In our culture, readers are accustomed to receiving information in little packages, "sound bites." The attention span is short, just as short for words as for speech. Producing a short *and* effective communication is not always easy. (Who was it who said, "If I had had more time I would have written you a shorter letter"?)

A few years ago, 3M compiled a list of suggestions for using less paper. The introduction said, in part, "By examining how we communicate, to whom and why, we can move towards less-paper-intensive communication, getting essential information (and nothing more) only to those people who need it." Here are some of 3M's ideas:

*In writing*

- Plan each communication. Define your purpose and audience. Before you begin a document, consider its life cycle from creation through disposal.
- Compose and review letters and other documents on the computer rather than on paper.
- Keep letters and memos to one page.

*In format and design*

- Redesign letterhead to condense headers (printed top material) and allow for narrower margins. Use up the old inventory, however.
- For routed information, put the distribution list cover sheet on the same page as the message. Examine publication formats and frequency. For example, send a two-sided, 8-1/2 by 11-inch newsletter rather than a four-page, fold-out version (cutting content accordingly). Consider publishing less often.
- Redesign and rewrite forms to reduce paperwork and form length. Eliminate unessential duplicates.

- For mailing, use the smallest envelope possible. Fold pieces to fit into standard business-size envelopes.
- Use the lightest weight paper possible.
- For large mailings, compile a prototype. Then combine pages to save paper and postage.

## *Through technology*

- Use the telephone if a verbal message will suffice.
- Use voice mail or e-mail for short messages.
- Make reports and data available on-line.

## *At the copy machine*

- Before running a large number of copies, do a one-page test of copier settings.
- Don't throw away slightly light or dark copies for internal use.
- Make double-sided copies, especially for large documents.
- Avoid making extra copies. Make extras later if you need them.
- Get training if you are unsure of copier features or hôw to use them.
- Prevent jams and toner problems by cleaning and servicing copiers regularly.
- Post a list of paper-saving copy ideas at every copier.

3M also suggests how management can support paper reduction. Flexibility and publicity are two keys. Departments should have the flexibility to examine their paper use and make changes, and to purchase paper-saving equipment. And the good things departments and employees do to reduce paper use should be publicized and promoted throughout the company.

## TAKING THE PAPER OUT OF PURCHASING

Everyone knows how a traditional, paper-based purchasing system works. The familiar steps:

1. Employee fills out a paper requisition form for what she or he wants.
2. Form passes upward for approval at one, two, or three levels of management, ultimately reaching the purchasing department.
3. Buyer calls supplier to confirm product is available and determine price.
4. Purchase order is produced and reviewed, and signed by the buyer.
5. Copies of the purchase order are mailed to the supplier and distributed to at least three internal departments, including purchasing, receiving, and accounts payable.
6. Supplier, upon receiving P.O., keys the data into processing system.
7. Ordered materials arrive at employee's company, alerting the accounts payable clerk to match data from all sources before making payment.

Elapsed time for this entire process? It can take as long as 3 weeks. Total cost? Hard to calculate for a single item, but industry statistics suggest a range of $75 to

$250 per paper-based requisition including all the staff time required to review and act on the request at every point up and down the line. Is there a better way? Many companies have discovered that there is, by moving all or part of purchasing into the realm of electronic commerce, on the Internet, replacing the preparation of paper documents with computerized forms and records that move from point to point instantaneously. Properly designed, e-commerce preserves all the checks and balances but eliminates a great deal of time—and, of course, paper. Typically, procurement on the Internet has three main steps:

1. Employee logs onto the company *intra*net, selects the purchasing page, and signs in with name and password. Purchasing page provides access to a predetermined list of commodities available from a preapproved number of suppliers. Employee browses the electronic catalog, selects the number of items needed, and fills in details about shipping and delivery.
2. Employee sends the order by e-mail direct to the supplier—with oversight copies routed automatically to the employee's department manager and the company controller.
3. Supplier verifies ordering data, ships direct to employee, who verifies receipt and authorizes payment.

Notice how virtually all the mid-section of the old purchasing system has been removed. Most of the checking and verifying steps that once occupied a platoon of clerks in the purchasing department have been handed over to the employee, as initiator, at the front end of the process; and to the supplier, as respondent, at the external terminus. To be sure, there are many safeguards built into the middle of the system—only a prescribed list of items that can be purchased, from a prescribed list of sources, at prenegotiated prices, for predetermined delivery addresses. And there is a most important overall prerequisite: a company *intra*net system that's been up and running for some time and is familiar and reliable. But given all these essentials, Internet purchasing can work just fine.

"The advantages of Internet-assisted procurement are as clear as an empty inbox," writes *Industry Week* magazine. "A Web-powered purchasing program can virtually eliminate the paper chase associated with traditional requisition and ordering procedures, as well as the labor costs and time lags that such efforts often create. It can also cut the need for in-house inventories through more timely product delivery." Case in point: National Semiconductor Corp., Santa Clara, California, was able to reduce its purchasing staff by 50%, reduce inventory by $1 million, and trim inventory holding costs $300,000 by adopting intranet–Internet procurement of maintenance, repair, and operations materials. On a smaller scale, the Gillette Co. transferred only 84 purchasing transactions from paper-based to computer-based but estimated it saved $1,260 in purchasing costs.

A cautionary note: e-commerce has been growing at a very rapid rate. One research company estimates it will account for $327 billion in transactions by 2002. Moving all that business between buyers and sellers at an acceptable pace is going to require huge new superhighways in cyberspace. Companies thinking about converting the purchasing function to an electronic system are well advised to assess the prospects for long-term gridlock on the Internet.

Purchasing on the Internet may be well and good for big, computer-savvy companies. But how can a small business do the same? It may, in fact, be impractical for a very small enterprise to acquire the software and Internet access essential to pursuing this sort of e-commerce. But with fax ordering and just-in-time, overnight delivery, there's no excuse for any business squirreling away more than a bare minimum inventory of office supplies, for example. The old office supply closet can be put to better use.

## Waste Prevention Attributes

It would be folly if an efficient purchasing system, operating essentially waste-free, delivered products that were inherently waste-full, especially if there were better choices available. Usually, there are. Certain kinds of office equipment, systems, and supplies have what could be termed "waste prevention attributes"—by design, they conserve materials and eliminate waste. Speaking on this topic at a meeting of the National Recycling Coalition, John Winter, senior research associate at Inform, Inc., offered the following list of such products:

*Equipment*

- Voice mail, e-mail, electronic bulletin boards
- Plain-paper fax machines
- Fax modems
- Computers and other equipment with modular features
- Computer software that allows employees to use templates to generate company letterhead and business forms on-line; send faxes from their PCs; peruse manuals, directories, and job postings on-line; file expense vouchers on-line; share information through networked computers
- Computer scanners
- CDs containing software documentation
- Optical discs for records storage
- Duplexing photocopiers
- Remanufactured toner cartridges
- Remanufactured photocopiers
- Refurbished or remanufactured furniture

*Supplies*

- Lower basis-weight paper
- Two-way billing envelopes
- Sturdy staplers, scissors, file holders, bookends
- Narrow-ruled notebooks
- Less toxic glues, pastes, glue sticks
- Colored pencils, crayons, or colored wax instead of solvent-based markers
- Perpetual calendars
- Erasable vinyl-coated boards for posters and signs
- Reusable interoffice routing envelopes
- Reusable mailbags and boxes

## Packaging and shipping

- Reusable shipping containers
- Reusable plastic crates instead of corrugated boxes
- Durable corrugated or plastic shipping pallets
- Reusable metal drums
- Paper shredder for making packing material from nonrecyclable wastepaper

## Food service

- Local fresh food
- Refillable condiment dispensers
- Bulk straw dispensers
- Bulk milk dispensers
- Reusable coffee filters
- Spun glass pads
- Washable rags
- Less-toxic cleaning fluids
- Cleaning supplies with concentrated refills
- Dry, concentrated dishwasher detergent in dispensers
- CFC-free refrigerators, freezers, and coolers
- Washable serviceware
- Reusable tablecloths, napkins, placemats, towels
- Mugs with no-spill lids
- Reusable cafeteria trays

## Maintenance and janitorial

- Washable furnace and air conditioner filters and filter frames
- Water-based paints
- Rechargeable batteries
- Washable rags, towels, mopheads, scrubbing pads
- Cloth-towel dispensers or air dryers
- Reusable vacuum cleaner bags
- Concentrated detergents, cleaning supplies, and disinfectants purchased in bulk
- Refillable pump spray bottles
- Less toxic polishes and waxes, disinfectants, degreasers, drain openers, window cleaners, caustic cleaners, toilet cleaners

Winter points out that source reduction products have common attributes. They are

- Durable
- Reusable
- Repairable
- Remanufactured
- Concentrated
- Less toxic
- Come with less packaging.

Purchasing policies can open the door to wider use of such products by getting rid of certain traditional purchasing language, such as the "all new material" clause that effectively blocks consideration of remanufactured products. "All solicitations should allow suppliers to provide acceptable substitutes," Winter says. The Minnesota Office of Environmental Assistance offers even more aggressive advice. In suggestions to business, the state agency recommends writing this sentence into bidding requests: "Preference will be given to products that create the least solid or hazardous waste while fulfilling the desired function."

Teaching a purchasing staff to think outside the usual guidelines takes perseverance, notes Marcia Deegler, an environmental purchasing trainer in Boston. Source reduction is not a visible attribute of a product, is not easy to define, and may be difficult to track and measure. But it can pay off big, she says, noting that the State of Massachusetts saved about $600,000 by purchasing remanufactured office panels, for example. At the Minnesota Pollution Control Agency, a group of 800 employees, an office paper reduction program produced these results:

- Converting the agency newsletter from distribution on paper to transmission over the intranet saved 400 pounds of paper a year—but not without a slump in readership as the changeover occurred: it took 2 years to recover.
- Converting from a paper timesheet to electronic reporting reduced paper use by 216 pounds a year.
- Replacing the mainframe computer with a local-area network (LAN) led to a 17% reduction in consumption of printer paper. This was a surprise—the expectation was that more paper would be consumed when local printing capability increased; in fact, e-mail came into play much more than expected.

Various changes like these enabled the agency to save $10,500 in paper purchases over 1 year. Cathy Berg-Moeger, an agency manager, points out that for initiatives like these to work, every employee must routinely use a PC. And to track results accurately, a central purchasing function is indispensable.

## Evaluating Two or More Products

When two or more products seem to meet general specifications for purchase but there is a question about which will contribute more to waste prevention, one way to choose is to make a detailed comparison. The Minnesota Office of Environmental Assistance suggests asking the following questions of each alternative:

1. *Cost.* What is the purchase cost of each product. Include price of extended warranty if planned.
2. *Warranted life.* What is the warranted life of the product?
3. *Durability.* What is the estimated life of the product in your application? This information may come from the manufacturer, maintenance records, or consumer publications. Is the product upgradable for a longer life?
4. *Repairability.* Is it cost-effective for the product to be refilled, remanufactured, or repaired? Does the product have interchangeable parts with other models currently in use?

5. *Quantity per year.* Based on life, what is the number of items needed for 1 year?
6. *Cost per year.* Cost of one unit multiplied by the number needed per year. The number needed per year may be a fraction of a whole number when the longevity of the product is greater than 1 year. Example: If product life is 4 years, then 25% of product life is used in 1 year.
7. *Weight per year.* What is the disposal weight of a product, including packaging? Multiply quantity per year by weight.
8. *Volume per year.* What is the disposal volume of the product, or a case of the product? Multiply quantity per year by volume.
9. *Disposal cost per year.* What is the cost to dispose of the product based on present service costs? If disposal cost actually would change, the amount should be added to Item 6 above.
10. *Toxicity.* What is the comparative toxicity of the product in use and disposal?
11. *Worker safety.* Including servicing and repair, what is the comparative worker safety involved in the use and disposal of the product? [N/A, Low, Medium, High]
12. *Labor.* What is the comparative labor expense of using the product? [N/A, Low, Medium, High] If the expense is known and there are staff changes, the amount should be added to Item 6.
13. *Other costs.* What is the comparative resource use of the product? [e.g., electricity, water, materials, etc.] If quantifiable, add to Item 6.
14. *Recyclability.* Is the product locally recyclable? Is its container locally recyclable? Does this affect cost? If so, add to Item 6.

## "The Company Store"

A case study in this chapter tells how an AT&T division in Florida cut the office supply budget simply by setting up a "Company Store"—a convenient place where departments dispose of surplus supplies such as file folders, three-ring binders, pens, pencils, clips, projector bulbs, etc., and other departments "shop" for their needs. This simple, free-access, untended, no-hassle, supply center saved AT&T an estimated $28,000 in office supply purchases over 1 year. Other companies have done the same thing. For obvious reasons, the purchasing department should play a role in setting up and monitoring an office supply exchange center. Too close supervision can kill the idea, however. Employees say they like the easy, quick access and lack of red tape.

Cinergy, the electric and gas utility based in Cincinnati, set up such a surplus store at one regional office when a corporate reorganization meant a shuffling of departments and functions. Such events have a way of unearthing little caches of office supplies in closets and storage cabinets as employees clear out of one place and move to another. To avoid disposing of perfectly useful office supplies (which happens more than office managers care to admit), Cinergy established "The Reuse-It Center." Following is the text of the announcement. Notice that employees are invited to take items for use by community organizations.

WELCOME TO THE CINERGY REUSE-IT CENTER
*What is the Reuse-It Center?*
The Reuse-It Center is a place where anyone in Cinergy can send unwanted office items that are in working condition and clean so that others may use them.

*What kind of items can go in the Center?*
Really, just about any unwanted item you would be throwing away in your office—or would just like to clean out of a cabinet or file—or things that have been there forever and are just taking up space. Like pens, pencils, staplers, dictionaries, file folders, binders, clips, magnets, lights, mouse pads, markers, portfolios, bookends—just about anything. Use your imagination!
*Why do we have the Reuse-It Center?*
This way, an item which may have gone in the trash can be used again by someone who needs it. It saves the company disposal costs, and valuable landfill space is not used. Also, if the item is reused within the company, we save the cost of buying a new one.
*Who can take items from the Center?*
Anyone. You may take any of the Center items for your own personal use, for a church, school, Little League, civic group, etc. And yes, even to use here at Cinergy!
*What do I have to do to take an item?*
Nothing. Just help yourself.
*If I have something to put in the Center, what do I do?*
You may send it to _____ in Environmental Services.
*Anything else?*
Yes. The Center is not staffed and we would appreciate it if you would keep it neat. If everybody does a little, nobody has to do a lot. By the way, the Center is always open!

## WASTE PREVENTION IN HOUSEKEEPING

Numerous specific methods for waste prevention in routine housekeeping are detailed below. But one of the most productive changes has to do with shifting responsibility from the housekeeping staff to the "tenants," so to speak. Some companies require office employees to play a more active role—by emptying their own waste baskets! That may seem odd: what's the connection to waste prevention? The link is through recycling. Offices that have set up rigorous paper recycling programs find that the volume of genuine waste in waste baskets, such as candy wrappers, dried-out ball pens, luncheon leftovers, and similar unrecyclable debris, plummets to the point where custodians are collecting very little material for disposal and quite a load of recyclable paper. Since it is a reasonable request to ask employees to tote their recyclables to a central collecting point, why not do the same with waste matter?

Such reasoning may not stand up to review in every office, but it does in some. One day at the Illinois Department of Natural Resources, employees were summoned to a meeting and told to bring along their individual wastebaskets. In exchange for the old cans, many of which were big enough to hold 14 to 16 gallons of stuff, each employee received two containers: a 12- by 14-inch recycling container made of cardboard, and a miniature waste basket measuring 5-1/2 inches tall and 5-1/4 inches wide. Staff members were told that starting immediately the janitorial staff would no longer collect recyclables and waste at individual desks. It was now each employee's responsibility to tote recyclables and trash to the central collection points in each department.

The idea, says recycling manager Jeri Knaus, is to force employees to think about what they throw away and what they recycle. The miniature waste can (Knaus

**Figure 1**  FORCED CHOICE. If the small container is reserved for true disposables and the large container for recyclable paper only, disposal volumes and cost plummet, companies say. (Photo: Frederik Richard)

calls it "this little bitty can") prevents indiscriminate behavior. Results? An audit at the 1,000-employee agency before introduction of minicans found about 2,000 pounds of white paper and 2,300 pounds of mixed paper per week. A few months after minicans arrived, the corresponding numbers were 3,600 and 3,000. White-paper recycling had increased 83%; mixed-paper recycling, 32%.

If recycling goes up, waste disposal must go down. Frederik Richard, a Canadian manufacturer of a miniature plastic wastebasket of 4-liter capacity, says that within 2 weeks of adopting a small container, companies typically experience a 15–20% reduction in the volume of material they produce for disposal. Companies that have modified their waste-hauling contract so that service is provided on an as-needed basis rather than daily or at some other prescribed frequency will benefit immediately from such a reduction. Companies that have not amended their contract will have an incentive to do so.

Besides cutting waste disposal expense, other economies often follow the adoption of miniature desk-side waste containers. Studies by the National Office Recycling Program USA and Environment Canada found the following, in a setting of 1,000 employees:

- *Use of waste can liners*
  Before—Changed three times a week
  After—Changed once a week

- *Janitorial costs for desk-side service*
  Before—2.5 janitors working 8 hours per day, 250 days
  After—Zero janitorial time

These studies also assume a 50% reduction in the volume of trash, which seems high for a company that has any sort of paper recycling program already in place. But the largest benefit by far is the release of janitorial time for assignment to other housekeeping tasks.

## Stop Unauthorized Dumping

Over the years, many employers have provided a fringe benefit that's not listed in staff benefit publications—free disposal of employees' home garbage! Especially if garbage dumpsters are within easy reach of parking lots, it may be tempting for employees to use the company as their household garbage removal service. There are two common remedies:

1. *Catch the culprit.* Since the unauthorized trash is usually bagged, it can be examined for any identifying marks, such as address labels on mail, to connect the garbage to an employee. AT&T found it had to do this at one of its New Jersey locations. Usually, a telephone call to the employee is enough to correct behavior.
2. *Remove the dumpster.* Indiana State University, in Terre Haute, no longer needed any large, open dumpsters after establishing a recycling program in office and classroom buildings. Trips to the landfill were cut by 70%. Where dumpsters have to be retained, moving them to a place that is inconvenient to street traffic helps avoid unauthorized use.

## WASTE PREVENTION IN FOOD SERVICE

Diane M. Mason and Carol W. Shanklin, faculty at the Kansas State University Department of Hotel, Restaurant, Institution Management and Dietetics, offer a large number of suggestions for preventing food waste, in their book *Environmental Issues Impacting Foodservice Operations* (Mason, D.M. and Shanklin, C.W. (1996). *Environmental Issues Impacting Foodservice Operations*. Manhattan, KS: Kansas State University. With permission.). Here are some key ideas:

1. Forecast menu production to decrease leftovers.
2. Encourage customers to take only what they will eat, and only the napkins they will use.
3. Allow customers to provide their own reusable take-out containers.
4. Provide reusable coasters instead of cocktail napkins.
5. Write menus that include what customers want. Form customer food committees, or let classes, clubs, and departments develop special diets.
6. Use permanent serving ware.
7. Purchase condiments in bulk.

8. Work with vendors to see what alternative packaging materials are available. Pouch packaging reduces the volume and weight of waste. However, steel cans may be a better choice if they are recycled.
9. To decrease food production waste, purchase products that are ready for consumption. This also may decrease labor costs.
10. Buy for simplicity and purchase the form of product needed. (Don't buy whole canned tomatoes if they are never left whole in recipes.)

Two of these suggestions—encouraging employees to provide their own, reusable drink containers, and using permanent serving ware—deserve a closer look, since they are so widely practiced by companies, or at least considered for adoption. The Minnesota Office of Environmental Assistance closely studied the waste prevention and cost advantages of these two measures in actual use at Itasca Medical Center, in Grand Rapids, Minnesota, a 143-bed institution employing 200 people.

## Reusable Cups

The hospital eliminated single-use styrofoam cups for staff use. It provided staff with high-quality, reusable plastic mugs embossed with the hospital logo. Staff are responsible for washing out their own mugs and must pay for a replacement if the mug is lost. Here are the numbers:

- Waste—At a consumption rate of about 1,000 single-use cups per week, avoided waste totalled 26 cubic yards per year. The reduction in waste-hauling charges ($6.26 per yard at the time of the study) would have been $162. However, no change was made in the hospital's contracted hauling volume.
- Cost—Assuming a useful life of 4 years for a mug, annual saving with mugs (at $1.35 each) over the purchase of single-use cups was calculated at $94 a year.

## Washable Serving Ware vs. Disposables

The cafeteria used reusable plates for most food service. Salads were an exception. But now they, too, have been converted, with the following results.

- Waste—By converting from 8-inch single-use plates to 8.5-inch reusable plates, the hospital avoided waste of 1,235 pounds a year. The corresponding reduction in disposal charges amounted to $225 a year. But as in the example of cups, above, this saving was not actualized, since there was no change in contracted hauling volume.
- Cost—After accounting for the cost of reusable plates and washing cost, the net saving was $2,126 a year.

As a further measure to reduce food waste, cafeteria customers pay by the ounce for salads they serve themselves.

## MAIL ROOM

The company mail room performs a number of vital services in the defense against waste. The first service, performed every business day, is controlling postal and freight charges and making certain that the company is using the least costly means of dispatching mail consistent with delivery requirements. But several non-postage services offer the additional potential for sizeable savings, including mail-piece design and addressing, and mailing list management.

Mailing room operations are so specialized by industry or profession, so subject to change, and so greatly influenced by geographic location that a detailed discussion is beyond the scope of this book. But an excellent reference on how to save money in the mail room is available free from the U.S. Postal Service at the USPS Web site, "www.usps.gov," or through the Rapid Information Bulletin Board, RIBBS. The Web site provides access to software titled Mail Flow Planning System, a program that can be downloaded to any IBM-compatible computer. (If you are starting from the home page, click on "Business." Next, click on "Mail Classification Reform." The MFPS is the first item. If you are using RIBBS, click on "Classification Reform." Next, click on "General Information." The MFPS is the next item.) The program covers, among other topics,

- Rate reductions
- Standards for earning automation discounts
- Savings on non-postage costs
- Mailpiece design and addressing

## Reducing "Junk" Mail

Advertising mail has acquired a bad reputation in recent years, though not entirely deserved. Shopping by mail saves significant amounts of time and travel expense. Still, the term "junk" mail sticks. For offices in which unwanted advertising mail is perceived to be a problem, there are effective solutions. One is described in detail by the Minnesota Office of Environmental Assistance:

> This reduction action was implemented in two courthouse departments to test its effectiveness. Here is how it worked: Anyone in the office receiving junk or duplicate mail deposits it in a collection box. Periodically, a staff person takes the mail and encloses a post card in the sender's preaddressed mailer. If this is not supplied, the person cuts off the portions containing addresses of the sender and recipient, attaches them to the card and mails it. The card reads: "To whom it may concern: In an effort to reduce our disposable waste, we request that you remove our name from your mailing list. Thank you."

This initiative succeeded in reducing unsolicited mail to about six pieces per week from previous counts between 75 and 100, a reduction of about 90%. The weight of waste avoided was calculated as 338 pounds per year. But did this reduction justify the cost? Without even considering the value of staff time on this chore, the added postal expense was $173 a year, far more than the cost to dispose of 338 pounds of material. Another approach to reducing the volume of advertising mail is to request the company name be removed from national mailing lists. The agency that handles these requests is the Direct Marketing Association, Mail Preference Service, P.O. Box 9008, Farmingdale, NY 11735.

### CASE STUDY: Millipore Corp.

Bedford, Massachusetts

*Moving toward a paperless office.*

Millipore Corp. manufactures and sells a range of purification products to the microelectronics, biopharmaceutical, and analytical laboratory markets. The company employs 3,200 people in seven plants and more than 30 subsidiary and sales offices. It does business in more than 100 countries. Through an aggressive program to prevent waste, the company has been steadily reducing its waste generation rate. In one recent year, while production rose 20%, waste declined by 5%.

Millipore has several programs aimed at reducing the use of paper:

- *Electronic mail.* For communication among the company's various offices, Millipore uses Lotus Notes groupware. It was introduced to this method by field staff, who found Lotus Notes the most convenient way to reach one another. From that point the method spread throughout the company, significantly reducing dependence on paper.
- *Double-sided copying.* Unless a special written request is made, all copying jobs submitted to the copy center are automatically duplexed. Purchases of white copy paper at headquarters declined 34% in the first year after adoption of the double-sided copying rule, which was implemented by the Facilities Department.
- *Catalogs on line.* The company publishes several product catalogs every year. A few years ago, the Corporate Communications and Marketing groups decided to place the catalog on the Internet as well, accommodating customers with on-line access. Customer feedback has been good, and the catalog print-order has been reduced.
- *Web page.* Millipore no longer distributes quarterly financial reports in print form. Instead, this data appears at the company web site, along with a wide variety of product information. The company estimates that by eliminating 60,000 printed financial reports it saves $25,000 a year.
- *Document control.* Millipore's document control system, developed as part of certification to ISO 9000 standards, is maintained electronically and can be accessed from any of the company's locations. Besides eliminating significant amounts of paper, the electronic system makes revisions available immediately, worldwide.

### Payback

The company believes its investments to date in paperless technologies, mainly the development of a presence on the Internet, have been compensated promptly through improved communication among employees, and between the company and its customers. It anticipates a continuing payback as growing amounts of paper are replaced by electronics.

### CASE STUDY: Silicon Graphics

Mountain View, California

*Replacing a paper-based purchasing system with electronic forms on the World Wide Web, a company eliminates 2-1/2 tons of paper and saves $2.1 million the second year.*

In 1994, Silicon Graphics, a computer manufacturer employing 11,000 people, began to seek alternatives to its purchasing system. Based almost entirely on paper forms, the system was straining under heavy loads. In a 2-year period, the number of purchasing transactions had increased 90%, to about 19,000 a year. The purchasing manager thought he would either have to hire more administrative staff or re-engineer the system. He chose to research opportunities in electronic commerce and, eventually, totally revamp the company system.

The project took more than a year and a half to complete. To begin, the purchasing group visited some 20 other companies to observe best practices in purchasing—a benchmark study. Concurrently, the company surveyed commercially available software for electronic transactions. Unable to locate software to meet its needs, Silicon Graphics decided to develop its own comprehensive purchasing system based on the World Wide Web and the company's internal network. Planning for the new system was informed by focus groups of employees and suppliers. The intent was to ensure that employees would find the new system more effective and that it would be compatible with supplier systems and procedures.

The Web-based system has led to a reduction in the number of purchasing steps from 15 to 3, and it has reduced order-filling time for many products from 3 weeks to 24 hours. A number of cumbersome and time-consuming steps have been eliminated, including the manual search of printed catalogs, preparation of a written request for approval, logging operations, form routing, supplier order forms, invoices, and receipts. Now, employees browse vendor-supplied electronic catalogs, input an electronic order form, route the form by e-mail for approval, then transmit it to the supplier by EDI—electronic data interchange.

Based on its experience, Silicon Graphics offers this advice to other companies interested in converting to a paperless purchasing system:

1. Study your paper-based system thoroughly. Know every step in the process, start to finish.
2. Look at the entire chain of people involved in purchasing, both customers and suppliers.
3. Listen to everyone's suggestions and concerns, and incorporate key ideas.
4. Don't be intimidated by initial start-up costs. Once the infrastructure is in place, the system can be expanded into other company departments where forms are processed, providing opportunities for additional savings.

### Payback

Although conversion of the purchasing system cost about $1 million, the company expected to reduce costs $200,000 the first year and $2.1 million the second, yielding payback within an acceptable time.

## CASE STUDY: FedEx

Memphis, Tennessee

*Send–return envelopes cut invoicing expense.*

Federal Express did not originally include a return envelope with its invoices to customers. But in 1992, in response to customer requests that a payment envelope be included, FedEx studied the overall cost and benefit of doing so. It was anticipated that most users of the return envelope would be smaller customers—those receiving no more than 11 to 12 pages of billing details in a typical invoice, with the entire mailing weighing no more than 2 ounces. For various reasons, it was believed that larger customers would not benefit from the inclusion of a payment envelope—they would probably use invoice processing and check mailing systems that could not readily accommodate the insertion of an envelope from an outside source. In fact, FedEx already was segregating its invoicing along these lines—invoices to larger customers were not mailed but rather delivered by FedEx couriers.

FedEx adopted as its standard billing mailer a No. 10, combination send–return envelope designed and manufactured by Tension Envelope Corp. FedEx mails about 125,000 invoices a day.

### Payback

FedEx did not disclose its envelope costs. However, the envelope manufacturer says an average user saves 15% immediately compared to the cost of purchasing separate mail-out and mail-back envelopes. In very large volumes, the savings would fall in the range of $2.25 to $2.50 per thousand envelopes, according to an industry source. Some users of send–return envelopes also report a positive effect on cash flow. Although this is difficult to explain, one theory is that the extra involvement of the recipient in following special instructions for opening a send–return envelope focuses immediate attention on the invoice. The send–return envelope also reduces "white mail"—payments from customers in an envelope other than the return envelope provided with the original invoice. Atlantic Electric Co., for example, had an 8% "white mail" rate when it used two separate envelopes for invoicing. The rate dropped to 4% with introduction of a send–return envelope, the company reported. (White mail is considered undesirable because the contents of envelopes are unknown and mail arriving this way requires special handling.)

### Additional Waste Prevented

1. The send–return envelope requires about 30% less paper to manufacture than two separate envelopes.
2. Because one envelope replaces two, warehousing and handling costs are cut about in half.
3. For companies that prepare invoices to customers using automatic inserting equipment, the single, dual-purpose envelope frees an inserter station for other uses.

**Figure 2**  PUSHING ENVELOPES. By lightweighting the Letter, FedEx saves millions. By billing customers in a send–return envelope, the express carrier speeds payment processing, besides cutting paper costs.

4. Since only a small strip of paper is removed to convert the mail-out envelope to the mail-back envelope, 90% of the envelope is returned. Thus if the company has a vigorous recycling program, most of the envelope paper is recovered.

The adoption of send–return envelopes has worked best when based on a "green" issue—an environmental concern—according to Tension Envelope Company. FedEx believes the message it sends to customers about waste prevention is an important justification of using a send–return envelope. Positive feedback from customers confirms this view, company officials say.

### *CASE STUDY: FedEx*

Memphis, Tennessee

*Pushing the envelope lighter.*

Federal Express picks up, sorts, and delivers more than 2.5 million letters and packages every day. FedEx is the world leader in overnight express delivery, with 1995 sales of $10 billion. As part of its service, FedEx provides shipping envelopes and boxes to its customers. Annual outlays for these supplies exceed $200 million. To control this significant cost of business, FedEx continuously examines its various containers and related shipping items in the search for ways to reduce size and weight without compromising integrity and appearance. FedEx works closely with its suppliers on this objective. "If I can improve my supplier's internal costs, he can pass the savings on to me," comments the senior quality assurance engineer.

The FedEx Letter—the familiar, glossy white envelope large enough to hold 30 pages of 8-1/2 by 11-inch correspondence without folding—is a good example. In 1981 the Letter was manufactured from solid bleached sulfate (SBS) stock, 100% virgin fiber, with a thickness of 20 points, or 20 thousandths of an inch. At that time, 20-point was the lightest stock available that met the company's requirements for envelope strength and printability. Today the paper industry (urged on by customers like FedEx) produces an acceptable 12-point stock, and FedEx uses that stock for its Letter, reducing paper requirements for this product alone by 40 percent. Considering that the company supplies more than 300 million FedEx Letter containers to customers per year, the overall reduction in cost is very significant.

### Payback

Since 1981, Federal Express has continuously reduced the thickness of the paperboard used to manufacture the 9-1/2 by 12-1/2 inch FedEx Letter envelope. Combined with savings from the lightweighting and redesign of other paper, paperboard, and polyethylene shipping containers, FedEx has cut its costs $20 million a year. Payback in outside supply cost is virtually immediate when a lighter weight is adopted. (Although the weight of the envelope would seem to be a factor in cargo capacity of FedEx airplanes, the company says cubic volume, not weight, is what limits aircraft capacity, and the cubic reduction of the FedEx Letter is insignificant in that regard.) FedEx is preparing to make a further change to the Letter envelope that will improve its recyclability—by eliminating the sheet of polyethylene film that is laminated to one side as a container for the airbill. "We don't want anything with our name on it going into the landfill," a company official says.

### CASE STUDY: AT&T Paradyne

Largo, Florida

*"Office Supply Depot" cuts costs $28,000 a year.*

The "Office Supply Depot" (OSD) at the AT&T division in Largo, Florida, is an unstaffed, self-service, mini-warehouse of common office supplies—a place where surplus items in one department can be picked up by other departments that have a need. The OSD team, an 11-member sub-team of the company's all-volunteer recycling and waste prevention group, takes care of collecting departmental donations and restocking the Depot's steel shelves (which also are surplus) with such items as binders, hanging folders, file folders, Rolodex cards, tabs, staples, staplers, staple pullers, highlighters, markers, pens, pencils, etc. "Shoppers"—AT&T employees from various departments—can browse and pick up supplies at any time. All they have to do is sign the log book indicating what they took.

One OSD team member was skeptical when he first heard the idea of a supply depot. "I imagined dirty used pens, broken desk trays, pencil nubs, and dried-out highlighters," he wrote in the AT&T Paradyne newsletter. To his surprise, departments scoured their closets and found large quantities of like-new but surplus office supplies. "The many unopened boxes received in the OSD should remind us to buy only what we need," the AT&T employee said.

"There are many benefits of using the OSD," another employee wrote. "First off, it saves money in your department budget. I check the Depot about once every 2 weeks. I've found such savers as refills for the Time Management System organizers, which saved almost $60. The Depot also saves me time. I can walk in, pick up what I need, sign the log, and I'm done. I don't have to look in the supply catalog, find the items, write a request, have my manager sign it, send in the request, and wait. I also find I order fewer supplies from the catalog because I can find what I need in the OSD."

Employees whose offices are some distance from the OSD can save a trip by checking what's in inventory at "company store" on the office data system.

## Payback

During the first 10 months of OSD operation, AT&T Paradyne employees reused

- More than 2,200 binders, valued at $2.45 each
- Nearly 9,500 hanging folders, $0.22 each
- More than 13,000 file folders, $0.16 each
- 1,000 Rolodex cards
- 560 pens and pencils
- 1,800 sets of dividers
- 1,000 binder clips
- Nearly 60 overhead projector bulbs

During the first year of OSD operations, the company estimated it saved a total $28,000 in avoided purchase of office supply items.

### CASE STUDY: Columbia University

New York City

*Five simple changes in purchasing–printing routines save more than $113,000 a year.*

Columbia University employs 9,000 faculty and staff and enrolls 17,000 students. The university computer and copier departments use approximately 5.5 million sheets of paper each year. As reported by the Waste Prevention Program of the New York City Department of Sanitation, Columbia made the following five changes to reduce the purchase and disposal of paper:

1. *Manuals.* By purchasing computer manuals on disks and copying them for distribution to staff, Columbia saves $3,000 to $4,000 per year formerly spent on the purchase of 8,000 hard copies. This practice also eliminates approximately 9,600 pounds of paper waste.
2. *Report distribution.* Many people at Columbia are on distribution lists for periodic reports. Each year, the production control group distributes a memo with each report asking the recipient to confirm whether the report is still needed. This practice has resulted in an estimated 8% reduction in computer paper usage, equal to 432,000 sheets (10 cubic yards) and an annual saving of $2,777 in paper costs.
3. *Covers.* Eliminating the printing of a header page (cover sheet) from 18 of 45 mainframe printers reduced annual paper usage by 75,000 sheets, saving $2,250 in paper and toner. (Headers could not be eliminated from the remaining printers because they serve too many different groups of employees.)
4. *Electronic data transfer.* The Office of University Publications prepares Columbia's employment ads—about 250 a year—for publication in New York City newspapers. Previously the ads were sent by mail. Now the artwork is sent by e-mail, saving $225 a year in paper and postage, and virtually eliminating delivery time.
5. *Duplex copying.* Duplex (double-sided) copying yields by far the biggest reduction in paper use and cost. Duplex copying is standard procedure at Columbia's two high-volume copy centers unless single-sided copying is specifically requested. To encourage two-sided copying, there is a per-page discount on a sliding scale relative to the number of copies made. About 60% of copy jobs are double-sided, saving 21 million sheets and $105,000 a year in paper costs.

## Payback

None of the changes listed above required any investment other than the time to develop new procedures and educate users to new routines. Aggregate savings per year: $113,252. Columbia University encourages its vendors to participate in the waste prevention program and make suggestions for improvement. (See copy of letter/questionnaire that Columbia sends to new vendors.)

## CASE STUDY: GTE TELOPS

### Westlake, Texas

*Refurbishing office carpet saves one-third to one-half the cost of new carpet.*

When Leo Scott, facility planning manager of GTE TELOPS, noted that the office carpet was showing signs of wear but not damage, he asked a carpet manufacturer if it was possible to simply wash, comb, cut, and redye the carpet. The carpet company conducted laboratory and field experiments to see if the suggestion from GTE was feasible. The result was the development of a proprietary process that extends the life of commercial modular carpet up to twice the normal span, according to the company. The process saves a significant amount of money compared to replacement and reduces the load on landfills.

**Columbia University**

**New York, New York**

**Dear Valued Supplier,**

In response to our concern regarding the amount of solid waste generated at our facility, we are currently looking into waste prevention measures. Some of the waste comes from deliveries and packing. We think that it is to our mutual benefit to look at options for reducing this waste wherever possible and that the result will be reduced costs for both of us.

Please let us know what initiatives your company has already taken to prevent waste and inform of us your future plans. Please fill out the attached questionnaire and return it as soon as possible.

Thank you in advance for your cooperation.

- - - - - -

SURVEY QUESTIONS

1. Identify any product changes that you have made that result in products being more durable, repairable, remanufactured, or less toxic.

2. Identify any product choices we could make that would result in products being more durable, repairable, or less toxic.

3. Identify any product for which you could eliminate unnecessary packaging or reuse packaging materials.

4. Identify ways in which we can help you to reduce or reuse the packaging of the products we purchase from you.

5. Identify ways in which you could reduce promotional material sent to us, such as catalogs or direct mail advertising.

6. Identify any products or packaging you sell which contain recycled content and state the percentage of post-consumer content.

- - - - - -

*(43 vendors responded to this request for suggestions.)*

**Figure 3**   VENDOR PROMPT. The subtle message to Columbia University's vendors is, we expect you to add value through waste prevention.

Sooner or later, most carpeting does find its way to a landfill. The refurbishing process delays the inevitable. Besides cleaning, the rejuvenation process may include retexturing and recoloring of carpet to blend with new office decor. The New England Co., Chevron Park, and Amoco also have used the proprietary process and report the same experience as GTE—cost savings from extended carpet life, avoided disposal cost of old carpet, and improved appearance of carpet.

To illustrate the volume of material and dollars involved, consider the GTE situation. Carpeted office space at the suburban Dallas location approximates 655,000 square feet. Using an average carpet weight of 1.5 pounds per square foot, the disposed weight would approximate 982,500 pounds, or about 491 tons. The cost to dispose of that weight of carpet at $25 a ton would be about $12,300.

GTE's original carpet varied in design. The carpet company developed a new design that blended well with wall coverings and furnishings.

In the case of Amoco (cited above), the company donated its refurbished carpeting to Colorado State University, out of concern not only about disposal cost but also the fact that old carpet manufactured from synthetics will remain virtually intact in a landfill for hundreds of years—a valuable resource buried. The refurbishing process resulted in a noticeable increase in pile height, the oil company reported.

## Payback

GTE estimates that payback occurred almost immediately. The calculation of savings was based on new carpet price of $25 to $35 per square yard, plus installation, compared to the reconditioning cost of $15 to $20 per square yard. Additional savings accrued from avoiding the cost of old carpet removal labor, and hauling and disposal cost.

## CASE STUDY: County of Alameda

Oakland, California

*Storing letterhead graphics in a laser printer cuts stationery expense.*

To conserve paper and reduce the cost of printing letterhead, memo forms, and other stock stationery items, the General Services Agency of the County of Alameda, California, stores printer templates in Word 6.0 for Windows. Templates include all routine letterhead typeset material as well as the county seal. To produce a letter, the operator selects the appropriate template from storage and then routinely keystrokes the contents, which are merged with the template and printed on blank paper. Template design can be changed at any time.

Word 6.0 uses the term *normal template,* meaning the default template that stores elements such as styles, AutoText, and macros. "When a document is based on a template, it contains all information to produce a specific type of document; for example, headers, footers, boiler text, and styles," quoting an instructional manual for Word 6.0.

## Payback

Telephone, fax, and e-mail numbers, and personnel names and titles, all are subject to frequent change. For years, organizations have simply gritted their teeth and scrapped outmoded letterhead and forms and printed new supplies when changes required it. By incorporating the county seal and other letterhead information in revisable, word processing templates, this waste is avoided. The main cost to adopt such a change in office procedure is staff time for designing templates and instructing workers in their use. Although no specific information about cost reduction is available, Alameda County GSA believes the use of laser-printer templates as a substitute for conventional forms printing is significant in terms of waste prevention and cost control.

## A Policy Statement for Office Staff

Many companies now include a statement on environmental responsibilities in the employee handbook. The best of these are not lofty pronouncements, but down to earth, very practical suggestions about how to minimize waste in everyday business affairs. The policy statement that follows appears in the AT&T employee handbook. Notice how it does not waste any words in getting to the point.

### *Waste Minimization*

In order to cut down on costs and decrease the amount of waste being disposed into landfills, AT&T has instituted a waste minimization program at all Company locations. The waste minimization program includes the following:

1. Employees are urged to print two-sided copies whenever possible to virtually cut the amount of papers used in half.
2. Jumbo rolls of toilet paper and paper toweling have been installed because they require much less packaging, and the toweling generates less paper waste.
3. The use of laser printers generates less paper waste in their printing process than the continuous feed printers.
4. Using electronic mail greatly cuts down on paper.
5. The packaging used for AT&T's products has been replaced with smaller, recyclable boxes.
6. Reuse binders, tabs, Pendaflex folders, manila file folders, paper clips, rubber bands, and other office supplies as many times as is practical. These items make up a large part of the waste stream and should not be disposed of prematurely.
7. Participate in the Environmental Mug Program (save 500 disposable cups per employee per year!).
8. Reuse interoffice envelopes until all of the address boxes are filled in.
9. Use disposable tableware *only* for "take out." Use non-disposable when eating in the cafeteria.
10. *Buy smart* when ordering supplies. Bear in mind whether the items can be reused or are easily recyclable once they are no longer usable. Also look for products that are made from recycled materials such as papers and plastics.
    • Before purchasing anything, check to see if the items are available used or as surplus stock.
    • Binders with plastic inserts on the cover and spine can be customized and reused easily.
    • Use mechanical pencils and refillable pens.
    • Avoid pre-printed binders, brown kraft and colored papers, carbon paper, or any papers coated with wax or plastic, including glossy papers. These items are very difficult if not impossible to recycle in some areas.
    • Remember, white paper is the most easily recyclable.
11. Avoid handouts at meetings. Use vu-graphs or make copies only upon request.

## Blast Off

For centuries, skyrockets have been launched skyward with gunpowder. Presumably, no one ever wondered whether there might be a better way to shoot them into the air—until The Walt Disney Company got on the case. Now, fireworks at Disney's Florida theme parks are sent soaring by compressed air, a waste-free propellant. It's the same glorious show as ever, with all the flash and boom, minus only the expense and mess of gunpowder.

The same kind of transformation is occurring in the "launching" of information by companies. For decades, the main propellant has been paper, billions of tons of paper. But in the new economy, where information technology itself accounts for as much as one third of economic growth, the old paper-based system is rapidly being replaced by electronics. Companies are learning that paper is not only unnecessary but prevents a timely response to matters moving at the modern speed of business. After reaping the many benefits of faster communication, it's almost anticlimactic to calculate the sizeable reduction in cost.

*The four pages of poster artwork that close this chapter were designed by the Minnesota Office of Environmental Assistance for free reproduction and use by business and industry in waste prevention campaigns. This artwork may be copied without permission and at no charge. However, the Minnesota source of the original artwork should be acknowledged in any use.*

POSTER ART

# It pays to repeat yourself.

Some copiers will do double-side automacally, while others have a manual feed slot. Know your copier — and cut paper use in half.

Reduce,
reuse, *then*
recycle.

Source: Minnesota Office of Environmental Assistance

Figure 4-a

# Reusing office supplies is a hot idea.

**Use reusable mugs and dinnerware to reduce waste and protect the environment.**

Source: Minnesota Office of Environmental Assistance

**Figure 4-b**

POSTER ART

 Reduce, reuse, *then* recycle.

Reduce, reuse, *then* recycle.

Reduce, reuse, *then* recycle.

Source: Minnesota Office of Environmental Assistance

Figure 4-c

<u>POSTER ART</u>

Figure 4-d

# Taking Control of Logistical Packaging: Pallets, Corrugated Boxes, Stretch Wrap

If you drew a simple picture showing how packaging moves through your business, it might look like this:

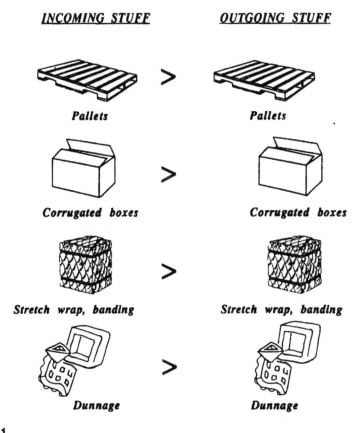

Figure 1

The two columns are identical, of course. And that's the point. If all incoming packaging on the left could be transformed into outgoing packaging on the right, the company's packaging and disposal budgets could be cut to zero, or virtually zero. At least there would be hardly anything left to dispose of at the end of the day. All arriving pallets loaded with raw materials would become departing pallets loaded with product. All incoming corrugated cartons would be transformed into outgoing corrugated packaging. Sound impossible? Companies are already doing it to one degree or another. Dow Corning, for example, in 1994–95 received 6,600 pounds of polystyrene packaging peanuts in incoming freight—and *dispatched* the same 6,600 pounds of peanuts as packaging for outbound freight. No additional packaging peanuts had to be purchased that year. Schumacher Electric, the company we met at the beginning of this book, found it could translate incoming corrugated cardboard boxes into packaging material for outbound shipments, saving $60,000 a year.

Besides passing packaging materials straight through the company, another strategy, very effective at curtailing the internal cost of packaging, is to build U-turns into the system. Thus, incoming pallets when unloaded are immediately U-turned back to where they came from, for reuse; outgoing products are packed in reusable containers that shuttle back and forth between producer and customers, a topic we examine at length in Chapter 5. This chapter explains how companies are not only building U-turns into packaging routes but applying intelligence in many other ways to cut the cost and waste associated particularly with pallets, corrugated cardboard boxes, cushioning materials, or *dunnage*, and stretch wrap.

## NEW ENVIRONMENT, NEW OPPORTUNITIES

Seldom has the time been better for companies to reexamine *logistical* packaging, meaning all the containers, pallets, and related items used for storage and transport of materials to and from the manufacturer, both raw materials delivered to factories and finished goods shipped to market. Logistical packaging, also called *transport packaging*, refers to the wrappings that keep manufactured goods secure and free of damage as they move through warehouses and distribution centers and over the road to destinations.

The seminal discussion of the new opportunities for producers in their use of logistical packaging appears in a paper authored by Diana Twede and published by the School of Forestry and Environmental Studies at Yale University. ("Less Waste on the Loading Dock: Competitive Strategy and the Reduction of Logistical Packaging Waste," September 1995. For further information over the Internet, http://www.yale.edu/pswp.) Twede is a professor in the School of Packaging at Michigan State University. She explains how deregulation of the transportation industry, along with the development of new packaging materials and rising disposal costs, have set the stage for buyers of logistical packaging materials to acquire more bargaining power with their suppliers than ever before.

"Until recently," Twede writes, "there has been little effort to reduce, rather than recycle, logistical packaging. Many costs associated with logistical packaging have gone relatively unexamined and unmanaged. Traditional logistical packaging designs have not changed for 80 years. The use of corrugated fiberboard boxes, steel drums, wooden crates and pallets has been institutionalized in the United States by transportation carriers, who maintain rules for 'acceptable' packaging."

Twede explains how the definitions of "acceptable" packaging, though widely thought to be government rules and regulations, actually were written years ago by the trucking and railroad industries, exercising authority conferred by the Interstate Commerce Commission. These old rules for packaging are precise descriptions of the materials and methods to be used in manufacturing logistical shipping containers. They reflect a close historical connection, Twede points out, between freight carriers and the Fiber Box Association, the trade group to which makers of corrugated boxes belong.

"For the last 80 years," Twede writes, "almost all approved packages have been made from corrugated fiberboard. The effect has been to create a virtual monopoly in the logistical packaging industry for corrugated fiberboard shipping container suppliers." Most of the packages, she adds, "use more material and are more expensive than they need to be."

This situation has changed, dramatically and rapidly. Transportation deregulation since 1980 has reduced the authority of truckers and railroads to prescribe logistical packaging. Among many other effects, deregulation has cleared the way for contract carriers to provide just-in-time delivery of full truckloads of freight, in dedicated service to a single manufacturer. Greater competition, especially among truckers, has shifted bargaining power, including the power to determine kinds of logistical packaging, from carriers to customers. But customers have been slow to respond, Twede writes. "Most firms are not accustomed to questioning the choice of logistical packaging, and they are not organized to facilitate packaging innovation."

Logistical packaging typically is made of the least expensive material that will do the job. Wooden crates were commonly used a century ago. Then the paper industry developed the corrugated fiberboard box. Reliable and cheaper than wood, this new material became the medium of choice for shippers. But today, Twede writes, "[t]he materials of logistical packaging are rapidly changing from fiberboard to plastic"—or, in the example of full-load transport of products from factory direct to user, changing from considerable packaging to very little at all.

## Milestones

Key developments during the transition from corrugated fiberboard boxes to other media include the following:

- *Plastic film bundling.* There are two main applications; shrink-wrapped bundles of case-lots of merchandise, such as food and pharmaceuticals; and stretch-wrapped larger items, including palletized goods. An example of shrink-bundling of case goods is the case study of Gerber Products Co. Instead of packaging glass jars of baby food in a corrugated honeycomb within a traditional corrugated box, the jars of product are tightly bundled without any separation—glass tight against glass—with only a bottom tray of corrugated. An example of stretch wrapping is given in the case study of Artec Manufacturing Division of Kimball International.

Instead of surrounding wall panels on all sides in a corrugated fiberboard box, only end-caps of corrugated are used, with plastic stretch film covering the remainder of the product. The principal advantages of plastic film packaging compared to corrugated fiberboard are reduced weight and cost; faster wrapping; better handling of packages in transit—a curious result of using transparent wrappings, revealing what's inside; and less packaging material to be disposed of at the receiving end.

- *Waste prevention mandates.* The landfill disposal "crisis" of the late 1980s and early 1990s was largely discredited a few years later. There is plenty of room for landfills and it is still comparatively cheap disposal space. But one legacy of the landfill scare lives on: state and federal objectives, and in a few cases state and local mandates, to reduce the volume of material sent to landfills and incinerators. Compared to economic motives, public goals and mandates are a distant second factor in waste prevention initiatives by U.S. industry. It's a different story in Europe, where packaging practices in several countries have been significantly influenced by Germany's take-back law concerning logistical packaging. More about this topic appears in Chapter 9.
- *Cost of waste disposal.* Disposal costs rose sharply in the late 1980s and early 1990s, reflecting less actual change in disposal capacity than, once again, the disposal crisis mentality of those years. In the later 1990s, disposal costs overall remained flat or declined, with some local exceptions. But industry had learned the lesson: the cost of waste disposal at *any* level is a measure of inefficiency. Significantly, the very success of waste prevention activities, including both recycling and source reduction, has flattened the waste generation growth rate despite steady increases in population and GNP.
- *Shifts in bargaining power.* The contrast between the thoroughly integrated, old and established corrugated fiberboard industry and the burgeoning, subdivided, highly competitive plastics industry shows up in the defensive tactics each has used. As part of its general response to preserving marketshare, the corrugated industry has orchestrated a long and highly successful campaign to promote the recycling of corrugated fiberboard containers. Most manufacturing companies of any size have a cardboard baler somewhere near the loading dock; often, the baler has been delivered at no charge by a paper broker. The price per ton of old corrugated cardboard is closely watched as an indicator of vitality of the recovered paper-fiber market. By comparison, the plastics industry appeared to have difficulty coming together in the early to mid-1990s to agree on a common message and marketing program.

About all this, Michigan State's Twede comments,

The corrugated industry's recycling effort has been an effective distraction defense against ecological concerns that favor packaging reduction. It distracts from the fact that corrugated fiberboard shipping containers use far more material than is generally required. The industry's recycling programs further institutionalize the use of fiberboard and provide another barrier to innovation.

While the corrugated fiberboard industry seems to speak with one voice, the plastics industry speaks with many, and the message is not consistent. Much of this is explained by the nature of plastics, which can be manufactured from a variety of specialized polymers for varying applications. In contrast, corrugated fiberboard is essentially the same material from one box to another, at least throughout the U.S. Although most plastic logistical containers and related shipping materials can be recycled, it's not nearly as easy to recover and recycle plastic compared to corrugated fiberboard. In addition, plastics for years have suffered from an "image" problem—manufactured from nonrenewable resources, often charged with being a big polluter, a main cause of litter, etc. In fact, most studies now indicate no advantage for paper fiber over plastic in terms of total load on the environment. One researcher, Frank Ackerman, writing in *Why Do We Recycle? Markets, Values, and Public Policy* (Island Press, 1997), reaches a more useful general conclusion. Concerning retail food and drink containers, Ackerman writes, "In almost every case, the lightest package, per unit of contents, is the one with the lowest impact on the environment .... Rather than legislating the choice of material or the required level of recyclability, it makes sense to adopt policies that encourage use of the lightest possible packaging for each product.

To the contrary, logistical packaging in the U.S. still depends primarily on the *heaviest* sort of packaging—wooden pallets and corrugated fiberboard cartons. They are so widely used for the transport of raw materials and component parts to factories, and finished goods from factories to stores, that economists from time to time have included the production rate of pallets and corrugated paper among indicators of national business vitality, like railroad carloadings. It takes immense quantities of wood and wood fiber to meet the demand for pallets and corrugated boxes. Regrettably, much of this logistical shipping material ultimately is put to a very low-value secondary use, often simply as the stuffing of a landfill or fuel for an incinerator. Because pallets and corrugated are such an obvious source of waste, many companies target them for early remedial action.

## SEARCHING FOR BETTER PALLETS

Pallets can be made of any durable material. The four most common materials in use today are wood, plastic, corrugated cardboard, and metal. Some pallets are built into the containers they support, such as the plastic and metal pallets that are integral to returnable–reusable containers (Chapter 5).

By far the most commonly used pallet is made of wood, including hardwood, softwood, and plywood. One fifth of all softwood and hardwood harvested in the U.S. goes into pallet production, an astoundingly large fraction. The pallet industry is the largest consumer of domestic hardwood lumber. It is a thriving industry. In 1980, 2,470 wooden pallet companies produced an average 112,000 pallets per year. In 1995, 3,793 pallet makers produced an average 254,000 units. Thus, over a recent 15-year period, the number of companies producing pallets grew 54% and average production grew 127%. Total wooden pallet production in the mid-1990s was estimated at 600 million units, or more than two pallets per U.S. resident, and the number keeps rising. California and Ohio have the largest number of pallet producers. The largest regional cluster of pallet makers is in the East North Central states of Wisconsin, Illinois, Michigan, Indiana, and Ohio. No coincidence, this area also is home to the largest concentration of manufacturing firms in the U.S.

**Figure 2**  UN-PALLETABLE. The U.S. pallet industry is the largest consumer of domestic hardwood. Expendable, one-trip pallets usually end up like this—a disposal expense. The search continues for better alternatives.

The most common type of wooden pallet is made of hardwood, contains about 18 board-feet of lumber, measures 48 × 40 inches, and weighs about 45 pounds in the flush-stringer, double-faced design. Purchased in quantity, such a pallet costs $4 to $6, a comparatively low cost that relates to life-expectancy. Because of extraordinary competitive pressure to keep pallet costs low, most pallet producers design their products as throwaways, with minimal material and expectations of little more than one trip as a platform for freight. While companies increasingly seek ways to reduce pallet waste, production of the expendable, one-trip pallet keeps rising—up 7% since 1985. Though most such pallets go to the junk pile in short order, a surprisingly large number do carry more than one load.

When disposal costs were low, it was common for companies to toss their broken or surplus pallets into the dumpster. Cheap disposal and cheap replacement was the status quo in the 1960s. Beginning in the 1970s and accelerating through the 1980s and 1990s, companies reexamined their pallet management strategies. There was ample incentive to do so: hardwood prices were climbing, landfill and incinerator fees were rising, and environmental legislation such as the Resource Conservation and Recovery Act underscored the importance of striking a better balance between disposal and recovery–reuse.

## Seeking the Perfect Wooden Pallet

Wooden pallets not only *move* freight; through human intervention they also *damage* freight. Over the years, industry has documented significant damage to products in transit caused at least in part by poorly made pallets. In the late 1980s, three organizations, the Food Marketing Institute, the Grocery Manufacturers of America (GMA), and the National-American Wholesale Grocers' Association, jointly sponsored a study aimed at reducing the estimated $2 billion a year loss of products while being transported on pallets. At the outset, the study group listed the qualities of the ideal wooden pallet. Clyde Witt, editor of *Material Handling Engineering,* ran the wish-list in his magazine in January 1990. Here it is:

1. Recyclable, one-way (possibly reusable) shipping platform. Preferably made from recycled material.
2. Exact 48 × 40-inch dimensions.
3. True four-way entry capability, that means able to be loaded and off-loaded with manual or powered equipment.
4. Overall height, 3 inches with 2.25-inch openings.
5. Solid top, nonslip surface with hand holes permitted.
6. Smooth, nonslip bottom-bearing surface with a minimum of 70 percent coverage and able to provide damage-free stacking.
7. Stackability: single, 3,000-pound units, five high.
8. Mechanization friendly.
9. Clearly identified for maximum unit load weight capacities: 1,000, 2,000, or 3,000 pounds.
10. Light weight: less than 35 pounds.
11. Ability to safely transport a unit load, damage-free, from point of manufacturing through any intermediate stops to customer's distribution center.
12. Sanitary and fire resistant, able to meet or exceed current industry practices.
13. No protruding fasteners.
14. Moisture resistant.
15. Nontoxic.
16. Adequate supply.
17. Cost: not more than $3!

The need for such a pallet was estimated conservatively at about 100 million units annually. Addressing an industry group a few years later, Witt reported lots of creative activity by the study group but no sighting yet of the ideal pallet. Designing in all the desirable features inevitably raised both cost and weight. But if such a one-way, expendable shipping platform seemed beyond reach, Witt said his own wish list would begin with a pallet strategy that already exists: leasing.

## Pallet Pools: Third-Party Rental Systems

The concept of renting pallets from a company that maintains a large pool of good-quality platforms has been around for many years. But the idea started to catch on with a significant number of U.S. users only in the 1990s. One of the largest rental systems has been established by Chep USA, an Australian-British owned company, with U.S. operations based in Orlando, Florida. In 1997, Chep had about 19 million of its distinctive blue pallets in circulation throughout the U.S., with significant market shares in the grocery, beverage, and produce industries. (The name Chep derives from initials of the full original name, Commonwealth Handling Equipment Pooling.)

Pallet leasing is a simple idea. Here is how it works:

1. The lessee—for example, a manufacturer or wholesale supplier who needs to move products from a production site to a distribution point—negotiates pallet leasing terms from a lessor, like Chep. Pallets usually are picked up from the lessor's terminal. Pallets are guaranteed very high quality. They are two or three times costlier to manufacture compared to a typical one-way pallet.
2. As long as the lessee's products are on the pallet, other parties involved in the transport chain, such as truckers, pay nothing for use of the pallet.
3. If the recipient of a leased pallet (other than the lessee) wants to use the pallet for some other purpose, the recipient becomes a lessee.
4. The final user in the transport chain is responsible for returning empty pallets to one of the lessor's depots. Lost pallets must be paid for in full—about $20 each.

Leasing of pallets offers the same overall advantages of leasing other kinds of office and industrial equipment—avoidance of investment in materials, maintenance, and management; a ready supply to meet demand; guarantee of quality, etc. For example, Chep's "Mark 55" pallet meets many of the performance objectives discussed above in the quest for the perfect pallet. When making proposals to prospective lessees, companies like Chep usually will observe the prospect's current operations to determine where costs are being incurred. The cost of labor associated with transport packaging—loading and unloading a truck trailer, for example—is sensitive to the kind of platform that products are placed upon, whether it's a slipsheet, an expendable, possibly broken pallet, or a uniform, high-quality pallet with four-way entry. Pallet lessors point out that it takes less time for experienced fork-lift operators to handle loads on identical pallets in good condition compared to mixed-quality pallets of various designs.

## Pallets in Grocery Distribution

The grocery industry probably is the largest user of pallets in the U.S. A 1985 survey, for example, found that grocery and related industries purchased 25% of total production that year of the standard 48 × 40-inch GMA (Grocery Manufacturers of America) solid wood reusable pallet, which is the most common size in use throughout all industry. Because the grocery industry is such a large user of pallets, and because the industry is very competitive and focused on cost reduction, it has an important influence on pallet design and management. In an article in the February 1996 issue of *Forest Products Journal,* authors Catherine Engle Scheerer, Robert J. Bush, and Cynthia D. West stated: "Many people involved in the grocery distribution industry consider the traditional pallet, along with its management system, to be a contributing factor to the perceived unacceptably high costs associated with materials handling."

Scheerer, Bush, and West conducted a survey of grocery distribution companies in 1994 to determine what kinds of pallets were in use, the perceived pros and cons of using various kinds, and a sense for the future direction of pallet management. Concerning present and projected use of the various types of pallets, the survey found the following percentage of companies using the listed types, with projected 1997 usage in parentheses:

*Solid wood pallets*—100% (94)
*Plastic pallets*—22% (37.2)
*Corrugated paperboard pallets*—0.5% (0.9)
*Wood composite [plywood] pallets*—0.5% (1.8)
*Other pallets*—5% (5.5)

(The *other* group includes Chep; FNPR—First National Pallet Rental; B-grade pallets, that is, recycled pallets different from GMA pallets but meeting certain standards; and metal cart pallets.) The significance of this survey finding is that solid wooden pallets are in universal use, but supplemented, to a significant and growing degree, by plastic pallets. Notice the big jump in anticipated use of plastic pallets between 1994 and 1997. Also notice that among alternatives to wooden pallets, plastic pallets are far out in front of the other choices.

Here are additional selected findings from the survey:

1. The decision to use a plastic pallet was often determined by the size of the order. Assuming that goods are received at a distribution point on wooden pallets, it is more efficient to fill a large order by pulling whole wooden pallets; conversely, it may be more efficient to fill smaller, less-than-whole-pallet orders by loading a plastic pallet.
2. Two different types of plastic pallets were used by survey respondents. One was a twin-sheet thermoformed, HDPE (high-density polyethylene) pallet, measuring 40 × 48 inches with true four-way entry, weighing 18 pounds, and with a dynamic load capacity of 3,000 pounds. The other was a structural foam molded HDPE pallet, again with true four-way entry but with a dynamic load capacity of 4,000 pounds.
3. Deflection under load was noted as a distinct disadvantage of plastic pallets. Users of twin sheet thermoformed pallets said they compensated for deflection by changing case-stacking patterns. (Other research indicates that the bending strength of wood is about three times greater than that of HDPE.)
4. Cost was the main reason cited for switching to plastic pallets. Respondents believed that repair costs would be lower and overall life-cycle longer than for wooden pallets. "Some believed that the plastic pallet (relative initial cost $23) would make 75 trips (distribution center to retail store and back) before needing to be repaired or scrapped," Scheerer, Bush, and West wrote. "In comparison, wood pallets (relative initial cost $4) were perceived to require repair, on average, after five trips." Thus, plastic pallets achieved a lower cost per trip.
5. Some respondents believed adopting plastic pallets for downstream shipment—from the distribution center—to the retail center complemented upstream use of a third-party wooden pallet leasing program. That is, both plastic pallets and leased wooden pallets place comparatively little maintenance demand on the user.
6. Some respondents believed plastic pallets, because they offer true four-way entry, improved the cube utilization aboard truck trailers. "With recent trailer width increases, true four-way entry allows more pallets to be loaded into a trailer with a pallet jack," the authors wrote.

7. Reduced worker compensation insurance costs was attributed to plastic pallets. Some users reported fewer back injuries to workers because plastic pallets weigh much less than wooden pallets.

Across all major product types, the survey found the most frequent use of plastic pallets in dry-goods warehouses. The authors speculated that this could have to do both with the concentrated use of pallets in those settings as well as with the environment. By comparison, they quoted the experience of plastic pallet users in frozen foods warehouses. Slipperiness was a problem—cases slipped off the pallet (a problem solved by placing a sheet of corrugated paperboard between the plastic pallet and the cases). "Another interviewee reported that their company was not using plastic pallets in the produce warehouse due to the possibility of organic debris accumulating in the cups (feet) of the plastic pallet which could create a sanitation problem," the authors stated. In other survey findings, corrugated paperboard pallets were used entirely with dry goods; composite wooden pallets were used only with produce.

## SPEQ-M Pallets

An alternative approach to lowering the per-trip cost is illustrated by the SPEQ (Specified Pallets, Engineered for Quality) program of the National Wooden Pallet and Container Association. This is an industry program to standardize pallet size and construction, with the objective of increasing pallet life. Pallet manufacturers licensed under the program agree to conform to certain prescriptive and performance standards. The so-called M-type pallets they manufacture should be good for at least 10 trips before they need repair. Such pallets bear the inscription SPEQ along with the supplier's licensee number. SPEQ-M pallets are in wide use among produce growers and shippers. According to the National Association of Perishable Agricultural Receivers (NAPAR), the per-trip cost of a SPEQ-M pallet is about $5, compared to $12–$13 per trip (all life-cycle costs included) for an expendable pallet. While the SPEQ-M pallet costs more initially—$9 to $10 per unit—the pallet will be bought back for $3, regardless of condition, by a licensed pallet manufacturer, thus also saving the user a disposal cost of $1.50 to $2.50 per unit.

## Recycling Wooden Pallets

Some old pallets are recycled. A survey in 1993 by the Center for Forest Products Marketing estimated that 66 million pallets were returned to pallet manufacturers for recycling or reuse. That sounds like a large number, but represents only 11% of total wooden pallet production at the time. Still, 90% of the pallets returned for recycling actually were reused, repaired, or dismantled and put to other uses. The repair and reuse of expendable pallets is an important factor in keeping pallet prices low.

## Kroger's Experience: Partnering with Vendors and Pallet Suppliers

The Kroger Co., the largest retail supermarket chain in the U.S., requires millions of pallets to keep merchandise moving to some 1,300 stores primarily in the Midwest and South. Gary Streepy, manager of pallets at Kroger's distribution center in Columbus, Ohio, recalls the early 1980s, when Kroger's pallet program focused on in-house repair and reuse. But this was no cost-saver, the company discovered—good pallets were being sacrificed to repair bad pallets. Kroger found it was not cost-effective to act as the principal supplier of pallets to its vendors.

The eventual solution at Kroger was to build a three-way partnership between the company, its principal pallet supplier, and its vendors. The objective was to transform pallets—traditionally considered a commodity and business *expense*—into an *asset*. It is a fairly simple change: Kroger maintains no pallets in inventory but negotiates with a pallet supplier (an Ohio company) to maintain a "bank" of pallets for use of Kroger and its vendors. The contract specifies pallet construction—GMA Class A, four-way entry, hardwood—and includes the cost of pallet repair. As part of its service, the pallet supplier agrees to deliver pallets to Kroger's vendors, or to hire other carriers for this job. In addition, the supplier allows Kroger's vendors to pick up pallets at the supplier's docks. As Streepy summarizes it, "We do a Chep [-style] program through Kroger." But the Kroger program was in place well before Chep arrived in the U.S., he adds.

Overall, Kroger controls pallet costs by transferring pallet storage and maintenance to a vendor but also by using the company's considerable purchasing power to manage the expense of providing transportation, *including* pallets. Streepy gives an example: if a vendor quotes $500 as the transportation component of a delivery of merchandise, Kroger, with much greater purchasing power, will say to the vendor, pay us the $500 and we'll take care of it. Then Kroger negotiates transportation for, say, $450, saving the difference. Kroger gains leverage with truckers in part by relieving them of pallet management responsibilities and in part by "completing the transportation route." Under other circumstances, Streepy explains, a trucker unloading merchandise at a distribution center would be required to haul away empty pallets equal to the number delivered. It's likely the pallets would be of mixed quality—some usable, some not. In any case, using or disposing of them would be the trucker's responsibility. But in the Kroger model, the trucker very likely leaves with an assignment to pick up at another Kroger vendor, and the pallets are all serviceable because they are maintained that way and go with the job. This saves the trucker from having to deal with pallets, and it keeps him from deadheading. Usually, Streepy says, Kroger only needs to cover the cost of fuel for the empty run to a vendor—a good deal for the grocery company.

For most "downstream" shipments—from the Kroger distribution center to its supermarkets—Kroger uses a twin-sheet, thermoformed, plastic pallet. As to durability, Kroger has had no problems with plastic pallets: only a few of the original 6,000-pallet inventory have been retired because of damage. But the company reports some familiar problems with its "retailer" pallet. The plastic surface does not grip loads the way a wooden pallet does, and the plastic pallet can slip out from under the product, an acute problem when working in the freezer, Streepy says. Also, because of the construction of a typical thermoformed pallet, with nine post-like feet molded into the bottom, it may be unwise to stack one pallet load on top of another, because the posts will poke into the material below. A bottom load of meat boxes, for example, might sustain damage from a plastic

**Figure 3**   PLASTIC PALLET. U.S. Postal Service is the largest user. Plastic weighs less, lasts
longer, but costs five times as much as wood. The key to low per-trip cost is getting
the pallet back—a "closed loop" system.

pallet stacked on top. Sanitation is another issue. Because of the way it is made,
a twin-sheet thermoformed pallet has many interior voids. Over time, even in
normal handling, the pallet shell may be breached, allowing water to enter. This
creates an environment where bacteria can grow, a very serious problem in the
food distribution business. Water trapped inside pallets also has an annoying way
of sloshing out on forklift drivers when pallets are being pulled off stacks for
loading. For the reasons cited, Kroger restricts use of plastic pallets to dry loads.
In addition, twin-sheet thermoformed pallets lack the stiffness required for storing
loads on racks. To compensate, merchandise handlers often will nest two plastic
pallets under a load, thus overcoming the stiffness problem but doubling the unit
cost of the pallet.

But many users of plastic pallets believe their advantages outweigh such problems. Long life is one clear advantage. At Kroger, Streepy reports a wooden pallet,
with repair, will survive five trips, cradle to grave. By comparison, plastic pallets
are very durable, and often are guaranteed for continuous use. Plastic pallets also
nest, conserving space—two to three times as many plastics can be stored in
equivalent space occupied by wooden pallets. The weight differences are significant.
Kroger's plastics weigh 12 pounds compared to 45 to 55 pounds for a typical
wooden pallet, addressing ergonomic issues. The U.S. Postal Service is believed
to be the largest user of plastic pallets in the world, with an inventory of 3 million
orange and black, 40 × 48-inch polyethylene cargo pallets for moving trays and
sacks of mail.

## Plywood Pallets

Plywood pallets typically consist of a solid top deck with a thickness of 23/32-inch (0.7187-inch); a bottom deck of the same thickness, with cutouts to accommodate pallet jacks; and nine solid blocks, equally spaced, separating and connecting the top and bottom decks and creating space for a forklift to enter from any side. The blocks may be made of solid wood, plywood, or plastic. Plywood pallets weigh about 70 pounds, or about 20 pounds more than typical board pallets. They can be safely racked with loads of 2,800 pounds.

Plywood pallets are stronger than board pallets, cost more initially, and last longer. The "Pallet Cost Estimator" (see illustration) was prepared by APA—The Engineered Wood Association, to demonstrate that even though the initial cost of a plywood pallet is more than a lumber pallet, the life-cycle cost is significantly less. The numbers are persuasive but apply only to certain industrial situations, defined generally as "closed" systems, where the pallets are under the user's control not only within the factory but throughout the external distribution system. If you buy a pallet for its reuse potential, you want to be sure it makes a round trip, and the best assurance is to bring it back in your own trucks. The food and beverage industries do a lot of that, using plywood pallets. The Stroh Brewing Co., for example, adopted plywood pallets in the 1970s and currently has more than 600,000 in circulation, moving product on its own trucks from five regional breweries to distribution points, with empty pallets returned on Stroh vehicles. The company reports that its plywood pallets last 7 to 10 years.

That lifespan appears to be confirmed by research. Procter & Gamble conducted an accelerated pallet aging test at one of its warehouses in 1989. As reported in a technical memo, 30 plywood pallets were subjected to 30 handling cycles simulating pallet handling over a 5-year period—and treated very roughly. The test ended before any of the pallets experienced substantial damage; for example, none of the decks needed replacement. P&G researchers projected that a plywood pallet had an expected mean time to deck replacement of 8.7 years. The corollaries were that 11.5% of pallets would need deck replacement each year; 5% would need block replacement; 21% would need block renailing. The most conservative estimated total cost of these repairs was $3.39 (1989) per pallet per year. By comparison, Chep, the pallet leasing company (and a participant in the P&G test) reported a history of replacing 1.3 deck boards three times per year on a standard board pallet. The cost of these repairs was (1.3 × 3 × $1.92 per board) $7.49 (1989) per pallet per year. Thus, maintenance of a plywood pallet amounted to 45% of the cost to maintain a well-built board pallet.

## ALTERNATIVES TO WOOD AND PLASTIC

The person who invents a pallet that is cheaper than wood but has all of wood's desirable characteristics stands to make a fortune. People continue to seek this pot of gold but so far have been rewarded only on a small scale in niche markets.

| | PLYWOOD STRINGER PALLET | LUMBER STRINGER PALLET | YOUR CURRENT PALLET |
|---|---|---|---|
| Typical Initial Pallet Price | $20.00 | $ 8.00 | _____ |
| Repair Cost (per life of pallet) | 1.65 | 2.75 | _____ |
| Stand-by Pallet Cost (based on number of pallets repaired monthly) | .18 | .24 | _____ |
| Cost of Disposal | .75 | .75 | _____ |
| Total Pallet Cost | $22.58 | $11.74 | _____ |
| Average Life of Pallets | 12 Years | 3 Years | _____ |
| Cost of Pallet per Year* | $ 1.88 | $ 3.91 | _____ |
| Annual Total Cost of Pallets (Estimated inventory of 40,000 pallets) | $75,200.00 | $156,400.00 | _____ |
| Annual cost savings between plywood stringer pallets and other types of pallets: | | − $75,200.00 <br> $81,200.00 <br> SAVINGS | − $75,200.00 <br> _____ <br> Your savings with plywood pallets. |

Form No. V690/September 1995/0010
©1995 APA – The Engineered Wood Association

**Figure 4**   COMPARISONS. The plywood industry presents these numbers to argue that original cost of a pallet may be misleading.

Ordinary board pallets supply 90% of the demand; alternatives to wood, such as
corrugated cardboard, metal, and plastic, probably account for less than 8% of the
market combined, and in special applications. Writing in *Resource Recycling*, Roger
M. Guttentag, a solid waste and recycling consultant, lists three main barriers to
market entry for non-wooden pallets:

1. *Cost.* With expendable pallets selling for $4 to $6 and the market ceiling at about
   $10 for the most highly engineered wooden pallet, alternatives to wood must
   surmount a formidable cost barrier. "Wooden pallet alternatives must achieve
   comparable or lower purchase costs as well as have similar or lower per-trip costs,"
   Guttentag writes. He notes that corrugated cardboard pallets "are often comparable
   in cost to wooden pallets and can perform the same functions."
2. *Performance.* Wood provides bending strength and stiffness without excessive
   weight. Wooden members also can be fastened together easily. "If the density of
   non-wooden components is heavier than wood, it will have an impact on shipping
   weights and costs," Guttentag writes.

**Figure 5** CORRUGATED PALLET. Key advantage of a paper-fiber pallet is it can be recycled, reducing disposal cost. Key disadvantage, little tolerance of moisture. Still, such pallets are used in transocean shipping.

3. *Recyclability.* "Most industries are a long way from agreeing on the use of standard-ized pallet products," Guttentag writes. "This means that, at least for the near future, an important advantage of wood pallets is the ease with which they can be dismantled and remanufactured into sizes that are in demand within a recycler's service region."

## Corrugated Paperboard Pallets

Pallets manufactured of corrugated paperboard have received a boost from the automobile industry. General Motors, for example, has specified delivery of loads under 500 pounds on corrugated pallets (see Appendix A). The advantage at the receiving end is that the pallet can be added to other paper for recycling, eliminating further handling and storage. And at GM, for example, receiving material on an easily recycled corrugated pallet is consistent with a policy on logistical packaging that specifically prohibits expendable wooden pallets. Because paper fiber loses integrity when wet, corrugated pallets must be used in a dry environment; obviously, they cannot be stored outdoors.

## Slip Sheets

Slip sheets are thin but tough platforms for sliding loads of cartoned goods from a factory or warehouse onto truck transport or ocean container and off again at the receiving end. Sheets measure 5 feet square and vary from one-eighth to one-quarter inch thick. They are manufactured of solid paper fiber, corrugated cardboard, or plastic—polypropylene (PP) or high density polyethylene (HDPE). One edge of the slip sheet is bent slightly upward to provide a grip for the fork-lift attachment necessary to slip loads across floors and docks. Slip sheets offer several important advantages compared to other platforms:

1. They weigh very little—2 to 5 pounds—reducing overall transport weight by a significant amount compared to a 50-pound wooden pallet, for example.
2. Because they are so comparatively thin, they create more cargo space, reducing unit shipping costs. Xerox, for example, learned that by shipping a certain product on slip sheets instead of pallets it was able to increase the load in a 40-foot-long container to 480 units from 400, a 17% increase.
3. They are the cheapest transport platforms—about $1 each in quantity.
4. Paper-fiber slip sheets are designed for a single, one-way use and can be disposed of at the receiving end by recycling with paper.

On the other hand, slip sheets require forklifts at both ends of the line to be fitted with a special attachment required to grip the tabs and move the load. Retrofitting a forklift can cost $6,000 to $8,000. Retraining the forklift operator adds more cost. Also, slip-sheet loads are best used for full loads that move direct from the factory loading point to the customer's unloading point. They are not designed for LTL, less-than-full-load shipments that move through an intermediate terminal. For a detailed discussion, see the case study of Home Depot, Inc., at the end of this chapter.

## STRETCH WRAP

Stretch wrap is used extensively to secure pallet-loads and to bind many kinds of merchandise for transport. The material is cheap and does a good job. The main problem is the expense of managing stretch wrap at the receiving end. It has virtually no further value as transport wrapping, though some value as recyclable scrap; but getting used stretch wrap into the recycling loop may be difficult. A marketable load must be clean, baled, and sizeable. Tenneco Packaging, for example, requires at least 5,000 pounds for free pick-up of stretch wrap, but only if it was originally purchased from Tenneco. Payment for the material doesn't begin until quantities reach the level of 30,000 to 35,000 pounds. For many companies, the cost of baling and storing stretch wrap for recycling far exceeds the cost of disposing of the material.

Strategies to eliminate stretch wrap or reduce its use fall into two general categories:

1. Minimizing use of stretch wrap by applying it with automatic equipment that stretches the material to its maximum.
2. Replacing stretch wrap with some other sort of cinch, either reusable or, if expendable, amounting to little waste compared to stretch wrap.

## Power-Driven Stretch Wrapping

Stretch wrap is made of LDPE—low-density polyethylene. The essential quality of this polymer is that it stretches, up to 300% of its length coming off the original roll. Stretching tensions the wrap and places a tighter grip on the load. This improves integrity of the load during transport *and conserves wrapping material*. Stretch wrapping a pallet-load by hand will consume at least two times as much material as wrapping with a power-driven machine, and, not incidentally, produce a less secure load. But stretch wrapping machines are expensive, ranging in price up to $70,000. It could take a long time to recover such an investment if the only saving is the reduction in wrapping material costs. Some of the other considerations are covered at the end of this chapter in the case study on the Charles Krug Winery, which converted from hand wrapping to machine wrapping.

Recently, a Colorado tomato grower applied a variation on power-stretching called "roping." In addition to stretching the film, the wrapping machine also compacts the 20-inch-wide film into a 4-inch band, creating a stronger wrap than using full-width. The company, Colorado Greenhouses, had been dispatching pallet loads of tomatoes using plastic banding to bind the cartons to the pallet. This method allowed some movement of cartons during a typical, jolting truck ride. Tomatoes arrived bruised from the trip. Stretch wrapping seemed a better method of securing the load but presented a problem: complete coverage would trap condensation and damage the produce. Roping was the solution, and at less cost per pallet. Banding had cost an average $0.48 per pallet; roping with stretch wrap cost $0.32 per pallet. Wrapping time also decreased, from 90 seconds per pallet with hand banding, to 30 seconds per pallet with machine applied stretch wrap in a roped configuration. This technique may be suitable for wrapping other kinds of perishable goods.

## Replacing Stretch Wrap

In Chapter 1 we learned how Traex/Menasha Corp. eliminated stretch wrap in warehouse operations by using large rubber bands to secure pallet loads. For a broad range of internal operations, such as moving product from the end of the assembly line into inventory before external shipment, reusable rubber bands can be a very effective waste prevention strategy. Another approach is to replace stretch wrap with a single-use wrap requiring less material. 3M recently introduced a stretchable tape that may cut materials cost and disposal expense in some stretch-wrap applications. The tape is 1-1/2 inches wide and comes in rolls. It can be applied with a hand-held dispenser or mounted in a stretch film wrapping machine. Application begins with a short length of unstretched tape, which is tacky before stretching and clings to cartons or merchandise. Once secured to a corner of the load, the tape can be stretched up to six times the unstretched length as it is wrapped around the load. Stretching the biaxially oriented tape nullifies its tackiness, so it will not tear labels or graphics; and also increases its tensile strength. 3M says the maximum tensile strength of its 4-mil-thick tape is 22 pounds; and the 6-mil tape,

40 pounds. Promotional literature shows the tape securing pallet-loads of uniform cartons, including produce; mixed loads with no more than a 3-inch overhang; and various unpalletized items. 3M says that 60 grams of its stretchable tape will do the work of 260 grams of stretch film.

## Glueing the Load Together

Some companies producing uniform corrugated cases or kraft bags of product have replaced physical wrappings such as film, tape, or banding, with a glue-like bonding agent that is sprayed on one side of each carton or bag as it comes off the packaging line immediately before stacking on the pallet. Thus, cartons or bags stick to one another and the load is held together from the inside. Manufacturers say this alternative costs half as much as stretch wrap and, of course, produces no waste at the receiving end. Palletizing adhesives characteristically have high shear strength (side to side) but low tensile strength (up and down). However, glues appear to have a limited application compared to external wrappings. They are best suited to automated spraying–palletizing of uniform carton or sack loads, with large areas of contact between cartons or sacks, and for direct final delivery of whole pallet loads.

## CORRUGATED BOXES

Diana Twede, the Michigan State University professor we met at the beginning of this chapter, has two general words of advice to purchasing agents concerning corrugated paperboard boxes. First, pay close attention to the market. Since the 1994–95 period, prices have moved through a wide range, peaking in 1995, bottoming in 1996–97. "Some less savvy customers are still getting charged higher prices," Twede says. Smaller companies that have dealt with only one or a limited number of carton suppliers over the years may be especially vulnerable. Twede's second suggestion is to consider replacing one-way corrugated containers with reusable containers (the topic of Chapter 5).

## Gapping the Flap

By shortening the major and minor flaps of a corrugated box, it's possible to trim a significant amount of material from the carton. For example, a 3-inch gap on the top and bottom of a box measuring 15 by 12 by 12 inches will save about 15% of corrugated material costs. The reduced cost of freight is a bonus. Gapping the flap may or may not be desirable, depending on how products inside the box are packaged, and the cost of adapting sealing systems to a gap will have to be analyzed for payback. But if safe delivery of the contents will not be compromised and changes in sealing cost are acceptable, this can be a quick and easy way to reduce materials. The Clorox Co. found that trimming a small amount of material from the flaps on cartons of bleach (see photo) eliminated 5 million pounds a year of corrugated carton purchases.

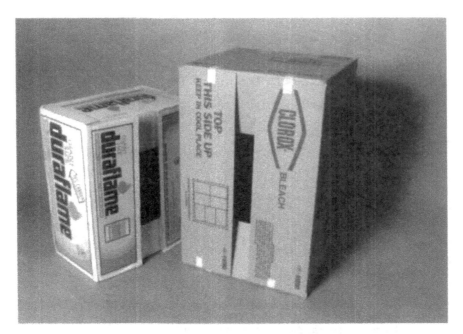

**Figure 6**  GAPPED FLAP. Depending on what's inside, it may be acceptable to trim the flaps of a corrugated carton. On a 12 × 12 × 15-inch box, a 3-inch gap top and bottom cuts material costs 15%.

## CASE STUDY: Cummins Engine Co.

Columbus, Indiana

*"Flipping" inbound pallets for use with outbound freight eliminates a major source of waste.*

Cummins Engine Co. is the world's largest manufacturer of diesel engines. Much incoming material and outgoing product moves on wooden pallets. Before adopting packaging standards in 1991, the company spent a significant amount of money to purchase new pallets and dispose of broken pallets. In 1991, Cummins published packaging standards for its worldwide operations. Among the general concepts that govern packaging management at Cummins is the following, under the heading "Packaging Viewed as Waste": "Since packaging is not a product feature that provides value to the end customer, we think of packaging as waste. Obviously not all packaging can be eliminated but major improvements are needed and possible if we view packaging as *variable cost* and *customers take the lead*." Cummins takes the lead on pallets by determining to reuse, or "flip," all possible incoming pallets as outbound pallets. To maximize the possibilities for reuse, the company prescribes the exact design and construction of acceptable incoming pallets, as follows:

> The pallet types and sizes shown are *the only standard pallet constructions* that will be used throughout the corporation. These styles have been chosen to accommodate business needs at lowest cost to the corporation since they will generally be reused.

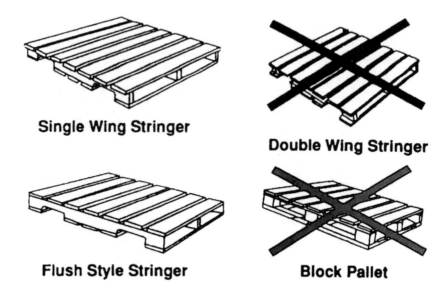

**Single Wing Stringer**

**Double Wing Stringer**

**Flush Style Stringer**

**Block Pallet**

**Figure 7**   PALLET SPECS. Cummins Engine Company tells suppliers what kind of pallets it will accept and reject. Banned pallets on right are more subject to breakdown because of the way they are designed and built.

*Pallet Requirements*

1. All pallets should have *4-way entry*. (Two-way entry acceptable with customer approval only.)
2. *Deck boards* must be 1/2-inch (13 mm) thickness minimum.
3. *Pallet stringers* shall have the minimum opening height of 3-1/2 inches (8.89 cm) on either end. The notched area on the sides should have a 2-1/2-inch (6.35 cm) overall opening height. All notches should be 9 inches (22.86 cm) long, with radial cut top corners, on 16- to 24-inch (40.64 cm × 60.96 cm) centers. All stringer pallets shall be double-faced, non-reversible. No single-faced pallets are permitted.
4. *Pallet strength* must be capable of bridging the 32-inch (81.48 cm) open distance between front support and rear support of tier racks.
5. Any *pallet material* other than wood needs customer approval prior to its use.
6. *Unacceptable pallet types:*
   • Double wing stringer pallet
   • Block pallet

Concerning packaging in general, Cummins cautions vendors to be on guard against items listed on this "High Cost Packaging Checklist":

• Steel banding
• Poly banding
• Triple-wall containers
• Oversize containers (non-standard)

- Excess interior dunnage
- Oversized and heavy-duty pallets (non-standard)
- Overkill on carton strength
- Die cut divider inserts

## Payback

The cost to "flip" pallets instead of disposing of them is minimal, and payback in terms of reduced disposal expense and new pallet cost is immediate. Cummins believes it has eliminated the major part of a variable packaging cost by adopting this pallet policy.

## CASE STUDY: Jefferson Smurfit Corp.

Litchfield, Illinois

*Streamlined case partition is lighter, stronger.*

Jefferson Smurfit Corp. (JSC) is the third-largest U.S. manufacturer of chipboard partitions, used to compartmentalize cardboard cartons of bottled goods. JSC also makes the largest multicell partition—one with 36 or more cells, mostly used for pharmaceuticals, cosmetics, and other small items. However, JSC was the smallest producer of the commodity partition—the 6-, 12-, and 24-cell partitions used extensively in the food and beverage industry.

To become more competitive in the commodity partition market, JSC decided to produce a new 12-cell partition. After 5 years of development, the company introduced the "Q Stack" partition. The glued, one-piece unit is manufactured of .037 chipboard, made of 100% recycled content. The new 12-cell partition replaces a slotted five-piece partition manufactured of .047 chipboard rollstock that was 20% heavier.

Although the one-piece unit is lighter in weight, it is stronger than the old partition. This is due primarily to the fact that it is a solid piece. Tests show that the .037 Q Stack partition has greater stacking strength than 175-pound test, B-flute corrugated cardboard commonly used to manufacture slotted partitions, and is comparable in strength to 150-pound, C-flute corrugated. Users also note that the old, slotted partition could shift and fall out of the shipping case, interrupting operations.

Q Stack is produced in six calipers from .028 to .040, in cell sizes from 1-3/4 to 4 inches, and heights from 5 to 15 inches. JSC lists the user benefits as follows:

1. Increased stacking strength
2. Improved load-bearing strength
3. Better sturdiness
4. Easier to insert into the corrugated shipping case
5. Reduced case, partition, and product damage
6. Reduced package weight

JSC says it has benefited from the new partition by:

1. Reducing material requirements by 20%
2. Reducing storage requirements
3. Reducing production costs by 10%
4. Reducing production scrap rate

## Payback

JSC made a significant investment in R&D to perfect the Q Stack and bring it to market. The company expects to recover its development costs over time with increased sales of the new partition. "There has never been anything like this available before," a JSC representative said. "This allows cost savings for our customers." Thus far the largest users have been glass bottle manufacturers.

### CASE STUDY: Corru-Fill/Corru-Shredder

Pompano Beach, Florida

*Old corrugated boxes become new packaging filler for outbound freight, saving $12,000.*

Norman Levine founded NDL Products, Inc. in 1974. The company, which Levine later sold, manufactures and distributes products for the sports and fitness industries. Raw materials and components arrive from suppliers in corrugated boxes; finished products are shipped to customers in new boxes—some 250 to 400 cartons a day.

Before 1992, incoming corrugated boxes were disposed of, and polystyrene packing peanuts were used to fill voids in outbound boxes of products. "I was tired of all this corrugated going to the dumpster," Levine told an interviewer at the time. "I also wanted to stop using the polystyrene peanut packing that was making a mess of our warehouse." The initial solution was to purchase a commercial cardboard shredder and chop the old corrugated boxes into pieces small enough to serve as packing material, replacing the polystyrene peanuts. But the pieces were too large, had a ragged look, and were full of small particles and paper dust. Customers complained. Levine and his associates then designed and built a new prototype which they called a Corru-Shredder, specifically to chop old cartons into uniform strips. The machine is 3 feet wide, 8 feet long, weighs about 400 pounds, and is mounted on wheels. It operates on 110 volts AC.

The operator flattens old cartons, then cuts them into strips, generally 4 to 6 inches wide, using a table saw. The width of this first cut determines the length of a finished piece of packing material. Next, the wide strips of corrugated are fed into the Corru-Shredder for slicing into uniform, 1/8-inch-wide strips. The finished material, called Corru-Fill, passes through a strong air current to remove dust, and it falls into a bulk container. As the container fills, the contents are sprayed with a biodegradable insecticide in an alcohol-based carrier. Machine maintenance includes daily oiling and occasional sharpening of the four shear blades contained within. On an

8-hour shift with two operators, the machine can produce about 1,500 pounds, or about 1,000 cubic feet, of Corru-Fill, according to Levine.

## Payback

By avoiding disposal of old cartons and the purchase of polystyrene peanuts, the company recovered its cost to build the prototype shredding machine in about 10 months. And since the machine was able to produce more packing material than the company needed, there was a surplus to sell in the local market. Assuming a free supply of old corrugated boxes, Corru-Fill is very competitive with polystyrene peanuts and is a superior packing material, Levine says, because it prevents the product from shifting during shipment, as explained below.

## Additional Waste Prevented

For equal volumes, chopped corrugated cardboard is significantly heavier than foam polystyrene. However, when Corru-Fill is dumped loose around products in cartons, the exposed corrugations on the thin strips of material cause adjacent pieces to catch on one another, forming a sort of lattice, Levine says. The many air spaces thus created tend to negate weight differences between corrugated and polystyrene, as well as reduce the required volume of packing material.

## CASE STUDY: Gerber Products Co.

Fremont, Michigan

*Shrink-wrapped bundle of baby food jars is just as secure as old-style carton with partitions—and 60% lighter.*

Listening to its customers, Gerber has heard growing concern in recent years about the cost of handling and disposing of logistical (distribution) packaging. Customer concerns of this sort are reviewed and acted upon by cooperative groups of company specialists representing all the affected areas, including:

- Purchasing
- Manufacturing
- Logistics
- Sales and marketing

The cost of packaging, though not the only concern, is extremely important. Disposable packaging traditionally has been made of the lowest-cost materials available—corrugated boxes came into wide use because they cost less than wood. But in many applications, corrugated is no longer the lowest-cost logistical packaging material.

Gerber was the first U.S. company to do away with corrugated partitions between glass jars, as well as the entire corrugated top, by adopting a glass-to-glass shrink bundle of low-density polyethylene (LDPE), secured to a bottom tray of corrugated.

Jars are confined so tightly they cannot rattle and break. (In preparation for the change, jars were redesigned to make them stronger, and surface coatings were developed to preserve integrity when glass contacted glass.) This change has significantly reduced the amount of packaging material that customers must dispose of. The reduction in package material by weight is about 60%, and the reduction in packaging materials cost is significant.

The new packing method also necessitated a change in the way Gerber received jars from the glass factory. The old way was for jars to be packed in a re-shipper—a partitioned, corrugated box used first to transport empty jars from glass factory to Gerber; and second, after washing, filling, sealing, and cooking, to transport them from Gerber to food stores. Today, jars arrive at Gerber stacked in layers on pallets, with a sheet of cardboard separating each layer. Jars are swept off the pallet layer by layer and fed into the washing area, and from there to filling, sealing, cooking, and packaging. Pallets and the cardboard separators are returned to the glass plant.

Gerber has retained a bottom tray of corrugated for several reasons. For one thing, it provides a hard surface for printing of various product information and machine scannable bar codes. Equally important, a corrugated tray provides some cushioning and protection for the glass jars during order picking, conveying, stacking, and loading.

### Payback

Reductions in materials expense enabled the company to recover conversion cost in what it described as an acceptably short period of time.

### CASE STUDY: Ethan Allen, Inc.

Danbury, Connecticut

*Company saves $1.15 million a year by shipping upholstery in film bags and small boxes.*

Five Ethan Allen factories package and ship upholstery—about 800 pieces per day at each site. The old method was to totally enclose each piece in a large corrugated carton. The new package is a proprietary system called LockPak. It works like this: a sofa, for example, is placed on a corrugated cardboard tray and covered by a polyethylene (PE) film composite bag. A corrugated end-cap is placed over one end. Then, the entire structure is encased in a PE shrink film bag, which is usually stapled under the tray bottom. Finally, the entire "Couch Pouch" is heat-shrunk either with hand-held heat guns or in a heat tunnel.

- Bags can be reused a number of times. Some retailers reuse them to wrap furniture for delivery to customers.

Compared to large, rectangular corrugated containers, Ethan Allen's new shipping package allows products to be nested, reducing storage and shipping space by

35% to 50%. Damage claims have been reduced. Packaging engineer Tom Lowery, as quoted in *Packaging Digest*: "We experienced a dramatic increase in production speed and our labor was reduced as much as half over boxing, which required a great deal of folding and stapling, plus the warehouse space needed to store raw materials."

Because package sizes are irregular, Ethan Allen developed new truck-loading techniques with the help of computer-assisted design (CAD). Loading time has been reduced by as much as 2 hours per truck. In summary, compared to the old container, the new package:

1. Allows products to be nested.
2. Enables manufacturer, handler, and customer to see contents.
3. Decreases damage claim rates (industry, 4%; Ethan Allen, 2%).
4. Reduces storage-shipping space up to 50%.
5. Reduces packaging materials storage space.
6. Increases production speed.
7. Reduces packaging labor by half.

## Payback

Savings resulting from reduced labor and materials, $750,000; savings in distribution costs, $400,000. Total: $1.15 million. Ethan Allen estimates it has reduced its consumption of corrugated cartons by a million pounds a year. The cost to adopt the new system was recovered in a year. The installation of a shrink tunnel at the company's largest plant reduced wrap time from the old 8–10 minutes using hand-held heat guns, to 1 minute in the heat tunnel.

## CASE STUDY: Artec Manufacturing, Division of Kimball International

Jasper, Indiana

*To ship modular wall panels, an "uncartoned" package works better than the old-style corrugated box.*

The Artec Division manufactures modular office wall panels of steel and fabric. A typical panel will measure 36 × 3 × 68 inches and weigh 70 pounds. The old method of shipping a panel was to wrap it in paper, protect the corners with corrugated blocks, and place the panel inside a double-wall corrugated box for shipment.

Technological advances in logistical packaging techniques plus a growing emphasis at Artec and its customers on reduction of shipping and disposal costs led to the development of the new method of packaging wall panels for shipment. Now, corrugated caps are placed on the corners of panels and then secured by wrapping in a horizontal stretch wrapper. The stretch film is 30 inches wide, cling-one-side. Two workers are required to lift panels onto the line and place the protective corrugated; a third worker operates the stretch-wrapping machine. Panels are moved by pallet to the truck trailer and there removed from the pallet and loaded individually, on edge, onto a padded floor. All panels for a job are stowed together.

The success of this shipment method owes much to adoption in the furniture industry of F-Packs (furniture packs), which eliminate most of the old corrugated box by substituting corner protection along with a total wrap in transparent film. This "clear-view" or "see-through" package has resulted in reduced damage to products in transit, which has been attributed to greater awareness of contents and care in handling.

### Payback

Artec invested approximately $100,000 in new machinery to convert to the use of an uncartoned, see-through package for its modular wall panels. Materials costs were reduced by 20%, and labor costs reduced by about 30%. The combination of all savings enabled the company to recover its investment in 12 months.

### Additional Waste Prevented

Artec calculates it has reduced the package weight of a modular wall panel by 18%. This translates into transportation fuel savings. Artec's customers have much less packaging material to dispose of than previously. Rising interest in the recovery and recycling of stretch film promises to divert growing quantities of this material from the landfill. One major U.S. chemical company now offers bins for consignee locations where large volumes of used stretch film can be gathered.

## CASE STUDY: Ford Electronics and Refrigeration Corporation, Electrical Fuel Handling Division

Bedford, Indiana

*Reusable shipping container eliminates 67.5 tons of corrugated cardboard, saves the company $150,000 per year.*

The Bedford plant manufactures an automobile fuel pump which requires rubber gaskets. Ford purchases the gaskets from Goshen Rubber Co., Inc., a factory located in Goshen, Indiana, approximately 190 miles north of Bedford. Gaskets are shipped from Goshen every other day by common carrier. During 1996, Ford purchased about 1.1 million gaskets.

Before adoption of a reusable shipping container in late 1995, gaskets were shipped in corrugated cardboard (CC) boxes fitted with CC grids to separate gaskets and avoid damage during transit. Boxes and grids were used only once and then thrown away. The company estimated it disposed of 67.5 tons of old cardboard a year at a cost of $100 per ton.

Development of a reusable container was a joint project initiated by the Lawrence County Solid Waste District, in which the Ford plant is located; in partnership with

**Figure 8**   REUSABLE SHIPPER. O-rings for fuel pumps move cleanly from the gasket plant to Ford in a reusable plastic tray. Rubber rings often stuck to disposable corrugated trays. Conversion saved Ford $150,000.

Ford, Goshen Rubber, and Robinson Industries, Coleman, Michigan, which designed and manufactured the reusable shipping tray that replaced the disposable container. In the interest of supporting a model industrial program to reduce the solid waste stream, the Indiana Department of Environmental Management provided a grant of $44,750 to cover about half of the cost to develop a reusable gasket shipper.

Besides the large volume of material disposed, the corrugated shipper had other drawbacks:

1. Corrugated fibers and paper separators often would stick to gaskets, causing rejects.
2. Projections molded into gaskets to align them properly in the fuel pump assembly often were damaged in transit.

Robinson Industries designed a reusable shipper of vacuum-formed, high density polyethylene (HDPE); 75% of the plastic content was recycled material acquired mostly from defective fuel tanks and damaged or obsolete dunnage from other Ford plants. Although trays are expected to last 7 years in continuous service, they may be made obsolete sooner by design changes. Trays that are damaged in the Goshen-to-Bedford loop will be recycled as new trays.

### Payback

Ford estimated that it recovered its investment of $42,050 in about 6 months.

*Other benefits*

1. The new shipper is designed to accommodate two different sizes of gasket, with easy visual identification.
2. Labor required to dispose of corrugated waste is permanently eliminated.
3. The Bedford initiative serves as a good example to other assembly plants and suppliers.

## CASE STUDY: Xerox Corp.

Webster, New York

*Factories and suppliers worldwide use—and reuse—nine standardized sizes of corrugated boxes.*

*Action*

Xerox Corp. manufactures and sells copiers and other office products worldwide. In recent years, competition has led Xerox to reevaluate operations and costs, including its supplier packaging program. Work began at the Webster, New York plant. In the past, this facility received component parts from more than 400 suppliers, each part packaged in its own unique box. This resulted in the use of thousands of different types of boxes and as many as 24 different pallet sizes.

To gain better control, Xerox developed a box reuse program (called the 88p311 Supplier Packaging Program). The company brought together packaging engineers from Xerox and its international suppliers to achieve a consensus on box style. Other participants in the planning process were quality control engineers, buyers, line engineers, assembly line workers, suppliers, and box makers. The program received final approval from senior manufacturing management.

The key features of the system:

1. Nine standard corrugated cardboard box sizes and two standard wooden pallet sizes have been adopted; suppliers are required to use them. A "Supplier Packaging Agreement Form" specifies how parts are to be delivered to Xerox facilities and describes under what circumstances exceptions are allowed.
2. A third-party handler manages the collection, sorting, and reselling of empty containers, eliminating the need to return containers to their point of origin. This is an "open loop" system.

Boxes and pallets can be used at any Xerox facility. They are designed to fit directly into designated positions on the assembly line, compatible with just-in-time delivery. As incoming shipments of parts are received and used, boxes are collapsed and stacked. The company then either reuses them itself to ship parts to other Xerox facilities or repair centers, or it sends boxes to the third-party handlers (one on each coast and one in the Midwest) who sort and resell the boxes to Xerox suppliers.

## Payback

Before adopting the new system, Xerox spent over $500,000 a year at Webster to send more than 4 million boxes to landfills. Now, standard boxes can be used for 60–80% of all incoming parts, and boxes average eight uses. Box usage has been reduced to 3.2 million units a year, a reduction of 2.4 million; $1.5 million has been saved in pallet disposal costs. Payback was almost immediate.

### Additional Waste Prevented

- Reduced freight cost because standard boxes and pallets "cube out" more efficiently.
- Reduced damage because boxes designed for reuse are sturdier; higher unit cost of sturdier boxes is negated by large-volume purchases and reuse.
- Reduced storage costs owing to just-in-time delivery.

All things considered, Xerox estimates that the new packaging program saves its manufacturing facilities between $2 million and $5 million a year.

## CASE STUDY: Home Depot, Inc.

Atlanta, Georgia

*By adopting plastic slip sheets, the company eliminates 1.8 million board pallets, cuts $660,000 of cost.*

### Action

Home Depot, the largest home improvement retailer in the U.S., has 423 stores nationwide. In early 1995, the company adopted a new policy requiring vendors to ship products to Home Depot stores on plastic slip sheets rather than wooden pallets. Some vendors were exempt because their products are too heavy or awkward, or are hand-loaded. After 2 years' experience, Home Depot made the mandatory slip-sheet policy voluntary, citing the high initial costs imposed on some vendors. But the company continues to encourage the use of slip sheets where possible and estimates that they are now used under 10% of incoming freight. Before the slip sheet policy took effect, more than 40% of shipments were on pallets.

Depending on volume, substituting slip sheets for pallets can benefit both the shipper and receiver of freight. Slip sheets:

1. Cost less than pallets.
2. Weigh much less, saving transport charges.
3. Are only a fraction the thickness of pallets, thus increasing cube and improving the efficiency of truck loading.

On the receiving end, slip sheets can be reused, typically 8 to 10 times. Home Depot reuses slip sheets arriving at its distribution centers to send products to its stores.

When the pull tabs rip and a slip sheet can no longer be used, it is sent to a plastics processor for grinding, melting, and remanufacturing. Slip sheets work best for merchandise with a heavy uniform case pack—products that "cube out" well. They have been used successfully by the food and beverage industry for more than two decades.

### Home Depot's Cost Analysis of Slip Sheets vs. Pallets
### (Per unit load @ 2,000 loads/year)

| Cost Category | Pallet | Slip Sheet |
| --- | --- | --- |
| Initial purchase | $4.00 | $1.80 |
| Special handling equipment | | $1.32 |
| *Quick fork attachment type push-pull device (assumptions):* | | |
| *$9,000 purchase/install* | | |
| *5-yr life with $1,000 total differential maintenance for push-pull attachment* | | |
| *Cost of money: @ 10% p.a. simple interest* | | |
| *Salvage value: 15% ($1,350)* | | |
| *2,000 loads per year* | | |
| *Total cost = ((9,000 − 1,350) + 900 @ 5 + 1,000) / 5 / 2000 loads* | | |
| Freight Differential: | $2.00 | $0.15 |
| *($10,000 cwt - pallets @ 40 lbs. vs. slip sheets @ 3 lbs.)* | | |
| *Assume pallet weight billed 50% of time* | | |
| Labor Differential: | | $0.43 |
| *Assume 2 minutes additional labor to stage & load slip-sheeted product @ $13.00/hr (fully loaded)* | | |
| **TOTAL COST per unit** | $6.00 | $3.70 |

## Payback

Home Depot realized an immediate reduction in freight costs and pallet disposal costs with the slip sheet program. The company developed the cost analysis (above) to show the relative cost structure of a typical medium-sized vendor shipping merchandise by pallets and slip sheets to Home Depot from a single location. The significant barrier to slip sheets is the initial capital investment (see "Special Handling Equipment"). The breakeven point is 3,600 loads. For a small vendor, such an investment may be prohibitive. Home Depot advises other companies interested in slip sheets to plan on a slow adoption process and expect some initial resistance from vendors.

## CASE STUDY: American National Can Co.

### Phoenix, Arizona

*Switching from wooden to plastic pallets for local delivery reduces costs and disposal volumes.*

American National Can (ANC) manufactures 12-ounce aluminum cans. Pallets are used to move newly formed cans to the company's primary customer in the same city, Pepsi-Cola. Empty pallets are returned to ANC on its own trucks. ANC had been using wooden pallets. But after experiencing maintenance costs and durability problems with wooden pallets, the advantages of switching to plastic became attractive. ANC adopted plastic pallets because:

1. Cleaning (removing wood chips, splinters) is no longer a problem.
2. Plastic pallets are made of recycled feedstock.
3. The brand of pallets acquired by ANC is guaranteed for 5 years.

ANC has experienced broken plastic pallets only through abnormal use, as when a pallet is struck by a forklift truck. Plastic pallets are assembled in two pieces, a top half and a bottom half that snap together and are fastened with plastic spindles. ANC reports that plastic pallets retain their shape after hard treatment that would crush a wooden pallet.

ANC encountered only one problem during the 2-year period to convert to plastic pallets: identification labels on pallets had to be repositioned to enable sensors to read them. Otherwise, ANC's experience can be summarized as follows:

|                | Wooden Pallets | Plastic Pallets |
|----------------|----------------|-----------------|
| Initial cost   | $12            | $50             |
| Weight         | 60 lbs.        | 50 lbs.         |
| Estimated life | 5 trips        | 2–3 years*      |
| Repair costs   | $2,000/month   | $0—No repair cost |
| Strength       | Same           | Same            |
| Broken pallets | Recyclable     | Recyclable      |
| Storage        | Some deterioration when stored outside | No apparent effect from outside storage |

\* Pallets had not been in use more than 2–3 years; manufacturer provides a 5-year warranty.

## Payback

Although the initial cost of plastic pallets is more than four times the cost of wooden pallets, ANC recovered its outlays in 2 years. A portion of the cost reduction derives from reduced disposal costs: wooden pallet waste amounted to about 2 tons per month. But the largest saving by far was eliminating labor costs of $2,000 per month for repairing wooden pallets. ANC's closed-loop control of its pallets—from can factory to Pepsi, then back to ANC on its own trucks—makes the system work.

## CASE STUDY: Pepsi-Cola Bottling of Phoenix

Phoenix, Arizona

*Pepsi also adopts plastic pallets within the bottling plant, with good results.*

As we learned in the case above, Pepsi-Cola Bottling of Phoenix receives cans from American National Can Company on plastic pallets, which are returned at once to the can plant. But Pepsi also has adopted plastic pallets to move product within the bottling plant. Pepsi has replaced half of its internal inventory of 44-inch by 56-inch wooden pallets with plastic pallets of the same dimension. Pepsi implemented the program after observing numerous problems associated with the delivery of can stock on wooden pallets. Slightly damaged or misshaped pallets sometimes splintered during automated unloading, interrupting production and causing substantial losses of new cans headed for the filling operation.

Pepsi asked its can suppliers to switch to plastic pallets. As a result, the plant manager reports: "Raw material losses resulting from pallet damage have fallen dramatically, labor to repair pallets for reuse has been reduced, and we have cut our production of wood waste by over 50 percent." He said the conversion from wooden to plastic pallets was easy and did not require any operational changes.

### Payback

Pepsi lists the following savings:

1. Labor cost to clean and repair damaged wooden pallets reduced by $3,000 a year
2. Wood waste generation cut in half
3. Production downtime (machine interruptions as a result of stoppage on the line stemming from defective pallets) cut from 10 hours a month to about half an hour
4. Useful life of pallets extended to at least 3 years or at least 30 trips, at least four times the performance of wooden pallets

It's important to note that Pepsi-Cola Bottling of Phoenix is using plastic pallets only within its own bottling plant. The company studied but decided not to use plastic pallets for shipment of product outside the plant, citing a number of reasons. According to the company, plastic pallets are

1. Much more costly than wooden pallets
2. Hard enough to retain and control within the internal system let alone permitting them outside
3. Not yet in wide use outside and therefore something of a curiosity and collector's item, aggravating the problem of getting plastic pallets back

### CASE STUDY: 3M Taiwan, Ltd.

Taipei, Taiwan, R.O.C.

*A simple die-cut piece of cardboard replaces an expensive PVC blister tray, cutting costs 86.3%.*

The 3M product in question is called a PCMCIA card—a modem part. The old package was a stack of PVC blister trays, with individual compartments molded to the shape of the card. Trays were stacked inside a corrugated shipping carton. The PVC mold was costly ($6,590). But there were two problems far more important than this initial cost:

1. Through normal vibration of the shipping process, cards were shaken loose from their compartments;
2. The PVC tray was so soft that it could be bent, exposing cards to damage.

In addition, the package consumed a lot of space for the number of cards it contained.

3M's new package for the PCMCIA card costs 86.3% less than its predecessor, holds much more product, prevents cards from shaking out of their slots, and has much higher compressive strength. It is manufactured from a single piece of die-cut corrugated cardboard, which can be reused or readily recycled.

### Payback

Replacement of the PVC blister tray with a simple corrugated box eliminated the difficult environmental problem of properly disposing of a special-purpose plastic container. After the one-time cost of designing the new container, conversion was relatively inexpensive; for example, the stamping die cost only $550. Besides reducing damage in transit, the new box also greatly increases cube utilization. 3M's new package for the PCMCIA card was awarded a Silver Star in the electronics category of the 1996 Package Awards Competition, Institute of Packaging Professionals.

### CASE STUDY: United Technologies (UT) Carrier

Indianapolis, Indiana

*Standard pallets are among a potpourri of waste prevention initiatives.*

UT Carrier is a major manufacturer of a variety of residential and commercial heating and air conditioning equipment. Carrier became fully aware of its potential for waste reduction after being introduced to the U.S. EPA "WasteWi$e" program through the Indiana Department of Environmental Management. The procedures recommended in EPA's *A Business Guide for Reducing Solid Waste* set in motion an overall management approach to waste prevention at UT Carrier.

One of the first initiatives was for the Purchasing Department to ask all suppliers, except those delivering steel coils, to convert to the use of standard GMA pallets (a 48 × 40-inch pallet first adopted by the Grocery Manufacturers Association). In 1995, this change made it possible to reuse 375 tons of pallets, avoiding a disposal cost of about $25,000. The company also worked with its insulation supplier (used in gas furnace and electric fan coil assemblies) to switch from delivery in nonreturnable cardboard boxes to wheeled steel carts, which are reusable indefinitely and move in a closed loop between supplier and buyer. The associated cost savings were not available.

Drums used for delivery of lubricant had previously been sent to a reconditioner at a cost of $10 each. But UT Carrier found that its oil supplier would take the drums for direct reuse at no charge. The saving on 120 drums a year is about $1,200. Similarly, empty 5-gallon buckets of silicone are being accepted at no cost by an outside company, reducing disposal costs about $2,000 a year for the 800–1,000 buckets used at UT Carrier.

The Purchasing Department has been gradually adding to its specifications the use of returnable/reusable containers for a wide variety of items, including electrical transformers, pressure switches, gas valves, prefabricated copper parts, blower motors, inducer motors, screws and fasteners, and burner systems. As selected parts have been successfully converted, increasing numbers of parts have been added to the system.

The largest reduction in materials use concerns corrugated cartons. By switching from full enclosures to simple top and bottom caps with stretch film around the middle, the company reports it has reduced purchase of cardboard by 2,100 tons, cutting costs by about $1 million.

## Payback

The company has not calculated a payback period for these various waste prevention initiatives. However, since none required any capital outlay but only procedural and purchasing changes, the cost savings are believed to have accrued at once.

## Additional Benefits

1. Vendors also have found that use of returnable containers reduces their overall costs.
2. Space for temporary storage of returnable–reusable containers was found within various work areas, avoiding the need to add plant capacity specifically for this purpose.

## CASE STUDY: Charles Krug Winery

### St. Helena, California

*Replacing a hand-wrapping process with semiautomatic equipment, winery reduces stretchwrap consumption 30%.*

The Charles Krug Winery has been operated by three generations of the Peter Mondavi family. Located in the heart of California's Napa Valley, the winery is considered one of the fastest growing in the nation—1996 sales grew 72% over the previous year.

For many years, the standard method of preparing shipments at Krug was to load cases five-high on pallets and to secure the load with stretch film, applied by hand. Typically, the winery gave the truck driver the stretch film and let him do the job, with assistance from Krug employees if time permitted. There were problems. Hand wrapping was time-consuming, taking as much as 15 minutes per pallet. In addition, because the film could not be stretched very tight manually, the wrapping had a tendency to loosen in transit, creating problems with unloading at the receiving end. Hand wrapping also could produce a messy-looking load.

As winery production increased in recent years, pallet wrapping began to become a time-consuming and labor-intensive bottleneck. After researching solutions, Krug decided to install a semiautomatic spiral stretch wrapper manufactured by Orion Packaging Systems, model H-66. It uses an 80-gauge linear low density polyethylene film. The machine's ability to prestretch the film helps to produce tighter, better-looking loads that are easy to unload. Loads now are stable enough that they can be unloaded with squeeze lifts.

The wrapping machine has two pallet-size turntables. While one is being loaded with cased products, the load on the other can be stretch-wrapped. Moving at 14 rpm, the machine applies two spiral wraps bottom to top and two more wraps top to bottom, plus one more pass around the bottom to secure the load. It takes about 3 minutes to wrap two loads. A company official said automated wrapping reduced by half an hour the time required to fully load a truck with 1,200 cases of wine (on pallets measuring 48 × 42 inches). Considering that daily shipments total about 5,000 cases, the reduction in loading time is significant, the company believes.

## Payback

The company declined to make public the size of its investment in automatic stretch-wrapping equipment. However, by stretching film to its full potential—300% of prestretch length—and by gaining better control of stretch film consumption, Krug has cut its annual expenditure for film from about $15,000 to about $9,000, or a reduction in film cost of about $500 a month. The winery feels well rewarded in other ways, most importantly the reduction in wrapping and truck-loading time and labor; improved integrity of pallet loads in transit; and the very positive feedback from customers concerning the appearance of shipments from Krug.

# CHAPTER 5

# Converting from Disposable to Returnable–Reusable Packaging

Returnable–reusable transport packaging is most evident in the auto industry, where mountains of reusable steel racks still define the landscape around many assembly plants. Car makers incorporated the return–reuse idea some years ago. It was a natural, with so many closed-loop, back-and-forth delivery systems feeding components from factories to the assembly point. Today, returnable–reusable packaging has been adopted by many industrial sectors, reflecting a wide range of concerns, including cost per product shipped, ordering–stocking, ergonomics, and just-in-time delivery, not to mention one of the original objectives—to cut the cost of disposing of single-use, one-way containers.

By far the most common one-way container is the corrugated fiberboard box. In Chapter 4 we examined some of the historical reasons for the primacy of corrugated boxes. The table below shows just how dominant this kind of packaging is —37%, more than twice as common as the ubiquitous beverage container, and used primarily for bulk shipments of food and beverages, and paper and printing. According to the U.S. Environmental Protection Agency, corrugated fiberboard boxes comprise 13.8% of municipal solid waste (MSW). On a per capita basis, U.S. commerce requires 217 pounds of corrugated boxes per person per year. Government statistics also show that old corrugated containers (OCC) are, by weight, the most recycled material found in MSW. The recovery rate for OCC ranges from 64% to 73%, a spread created by different methods of determining the base.

| Leading Kinds of Packaging in the Municipal Solid Waste Stream | |
|---|---|
| Corrugated boxes | 37% |
| Beverage containers | 17% |
| Food & other containers | 13% |
| Wood packaging | 12% |
| Folding cartons | 7% |
| Bags/sacks | 5% |
| Other | 9% |

*Source:* EPA.

Despite very impressive rates of recovery and reuse, corrugated boxes are the main area for reform of transport packaging. That's not surprising given their extremely wide use. Surely there are unexplored opportunities to replace corrugated boxes with other materials, and to cut costs without compromising the security of shipments. Case studies in this chapter show how companies have found such opportunities and capitalized on them.

## HOW TO THINK ABOUT RETURNABLE–REUSABLE CONTAINERS

In a 1994 publication titled *Delivering the Goods: Benefits of Reusable Shipping Containers* (Inform, Inc., New York), author David Saphire identifies four features of a product-transport system that indicate the desirability of adopting reusable containers:

1. Short distribution distances
2. Frequent deliveries
3. Small number of parties [in the system]
4. Company-owned or "dedicated" distribution vehicles

We have already seen cases illustrating these principles—Ford's adoption of a reusable plastic tray for frequent delivery of fuel-system gaskets from an in-state source; American Can's conversion from expendable wooden pallets to reusable plastic pallets for across-town deliveries to customer Pepsi-Cola, in trucks dedicated solely to that purpose. In a reusable-container system,

- *Short distance* minimizes the cost and time required to haul empty containers back to the source.
- *Frequency of delivery* indicates rapid use of the material delivered ("inventory turn") and continuous production of empty containers for backhaul, thus reducing the total number of containers required to keep the pipeline full of containers.
- *Small number of participants in the system* enhances control; two is ideal, one at each end; three is manageable (and illustrated by the Xerox case in Chapter 4).
- *Company-owned vehicles,* or third-party vehicles dedicated solely to carrying the products of one company between two points, may have the effect of reducing direct charges for the return of containers (though by how much is debatable since labor, fuel, maintenance, etc., all are required for the return trip).

### A Systematic, Seven-Step Approach

The Minnesota Office of Environmental Assistance believes that reusable packaging delivers the best return on investment when it emerges from a careful, comprehensive evaluation of need. The state agency recommends a systematic, seven-step approach, which is excerpted below.

1. *Organize a team.* Individuals interested in reducing packaging waste make effective team members, especially people with good communication skills from depart-

ments affected by transport packaging changes. Purchasing and marketing staff, in addition to those who handle containers, are important. A facilitator from the team makes sure that progress, assignments, and reports are distributed, and that communication remains open.

- The team is the link between management and the "packaging handlers" to see that good ideas get support.
- The team prioritizes ideas, with those that have the greatest potential for cost savings at the top. However, packaging performance standards must be known first. Employees who unpack materials, truckers, people receiving the goods you ship, suppliers, and packaging engineers can identify performance needs.
- The team evaluates cost and waste benefits for each packaging change.

2. *Ask for suggestions.* People who handle packaging will have good ideas about how to reduce it. Make it easy and safe to offer suggestions. The flow of materials into, through, and out of your facility and assembly line is important to examine. Questions to ask include the following:

- Can this package be eliminated? Individually contained products can sometimes be bulked or concentrated.
- Can this package be minimized? Thinner, lighter, or less packaging may do the job.
- Can you use a recyclable package made with recycled content? Even minimized or reusable packages should be recyclable.

Encourage everyone, including fork-lift operators and truck drivers, custodial staff and the boss, to put their ideas into suggestion boxes. Signed suggestions allow recognition. Work cell or unit meetings and e-mail are other ways to gather ideas. Don't forget to ask for ideas outside your facility. Suppliers who participate with you in just-in-time delivery systems are often excellent candidates for reusable containers.

3. *Determine the best type of container.* Many types of source-reduced or durable packaging are available—bags, boxes, bins, totes, pails, drums, racks, and pallets. Packaging suppliers will gladly supply samples and information. Reusable containers should stack, nest, or collapse for back-hauling; they should wash easily. Whatever the container type and material, it should make handling, emptying, and filling easier for everyone. Gather samples.

- Preliminary cost–benefit information can be obtained at this point to determine which, if any, merit a test run.
- Some samples may be clear winners; others may require help from a packaging engineer. Source-reduced packaging is effective if product damage does not increase. Reusable packaging requires closed-loop distribution.

4. *Test sample containers.* Walk a sample through the distribution system to see how the container works. Write down the opinions of users. Don't expect this to go smoothly, especially if your first effort involves distribution outside the facility. Changes of any kind are rarely accepted without resistance. Use the feedback you get to modify the container or the distribution system to maximize efficiency. Address everyone's concerns before proceeding, and keep management informed. If issues are too complex or controversial, choose another suggestion and walk it through its distribution system. Go with the first suggestion that works.

5. *Document final changes in cost and waste.* Keep careful records of cost changes. Staff time decreases for purchasing, stocking, assembling, opening, and disposing of single-use containers. Disposal costs go down. Product damage typically decreases. Reuse often improves working relationships within the facility, with suppliers, and with the community as environmental benefits are made known. Write up the issues and benefits, including the payback period.

- Use the most cost-effective container but stay flexible. A nestable, corrugated *plastic* container may be best for long-term savings, but piloting with a reusable corrugated *cardboard* container may help uncover communication problems unrelated to packaging.

6. *Change one package at a time.* Tests will show which packaging change has the best chance of success. Changing one package at a time allows people to adjust to the "reuse, reduce" principle and lets modifications be made without disrupting productivity. Announcing the decrease in waste generation helps people understand the environmental effects of their actions. If steps 1 through 4 were done well, this process can go smoothly.

7. *Establish a feedback method.* After a packaging change, team members should ask everyone involved—suppliers, assembly-line workers, maintenance, truck and fork-lift drivers—for feedback frequently so that small problems don't become large ones. Keep management informed. Use staff meetings, newsletters, posters, and e-mail to inform employees. Report to local citizen groups. Reward personnel who demonstrate commitment to reduce packaging waste and costs.

(The Minnesota Office of Environmental Assistance has assembled a directory of suppliers of reusable transport packaging. Copies may be requested from the agency: 612-296-3417 or 800-657-3843.)

## Words of Caution

"Critics of reusable programs say that purchasers may not always be doing their homework," Rick LeBlanc writes in *Pallet Enterprise*. He observes that the decision to adopt a returnable container program may be unduly influenced by such things as environmental correctness, without regard, or in spite of, analyses of cost and efficiency. In a report titled *The Hidden Costs of Reusable Containers,* Stone Container Co., a major U.S. producer of corrugated boxes, lists a number of factors that may escape attention when companies consider replacing expendable containers with reusables, including

- *Loss of cube efficiency* because of adopting a limited number of reusable totes and failing to make the most efficient match with the cube of freight; and building air into the load by using nesting boxes with angled sidewalls.
- *Increased weight* owing to the materials and design of reusable containers. A reusable plastic tote can weigh up to five times as much as the equivalent corrugated box.

## THE JOHN DEERE STORY

One of the largest corporate conversions to returnable–reusable containers during the 1990s occurred at the John Deere & Co. plant in Horicon, Wisconsin. The process that Deere followed and the results the company achieved in waste prevention stand as models for U.S. industry. We are indebted to Steve Buchholz, manager of the Deere project from its inception in 1991, for the details that follow.

Deere's plant in Horicon employs 1,500 people and produces ride-on tractors for lawn mowing and grounds maintenance. More than 100 suppliers ship components to the Wisconsin plant. Before adoption of the returnable container program, virtually all freight arrived in one-way, expendable containers. Most of this was corrugated boxes. Some could be baled and recycled. But a significant amount of the corrugated, at least one-third, was wax coated or contaminated with oil and therefore not recyclable. It had to be landfilled. Buchholz recalls the time when four loading-dock doors were dedicated to moving old corrugated containers out of the plant. In 1988, the company sent more than 1,500 40-cubic-yard compactor loads of trash to the landfill. Most of the trash was transport packaging, including wooden crates.

The company had been tracking its disposal costs and was concerned about them. In the late 1980s, projections indicated that landfill disposal fees would be one of the fastest rising costs of doing business. Like many companies at the time, Deere considered the disposal expense simply part of doing business. However, a revolutionary change in thinking began at Deere in mid-1990, shortly after the Wisconsin legislature passed a law prohibiting the landfilling or burning of most containers, effective January 1995. As Buchholz observes, companies historically have viewed such enactments as simply more work and expense for the private sector. But a team of Deere employees viewed things differently. They saw the new law as a challenge to reinvent the company's transport packaging system—by adopting returnable containers. They would measure success by reaching several challenging goals: eliminating waste, streamlining material handling, improving the work environment, and reducing costs.

## Team Approach

The Horicon plant created what was called the Corporate Returnable Container Team, with members from purchasing/procurement, environmental engineering, and other departments with an interest in the project. The team's first task was to gather information about the use of returnable container programs in other industrial settings, beginning with visits to automobile assembly plants. Auto makers were the recognized leaders in developing returnable container systems. Expert advice also was sought from materials handling companies, consultants, and other outsiders.

Then the team turned to perhaps the most important stakeholders in the entire project—Deere employees and the suppliers who would be directly affected by the adoption of returnable–reusable containers. Armed with much information and many points of view, the Deere team at Horicon was ready to begin looking at specific containers for a pilot program. Elsewhere in the company, people were watching: word of the pioneering work at Horicon had spread to Deere's 10 other manufacturing sites in the middlewestern U.S. and Ontario. If returnables worked well at the Horicon facility, the program could be rolled out to all locations.

## Selecting Containers

To identify the best containers for the Horicon plant, 100 Deere employees—10 salaried, 90 hourly—spent 2 months closely examining the various choices. They formed a tough jury, taking containers out to the assembly lines and handling them

**Figure 1**  RETURNABLE–REUSABLE. Plastic totes are widely used for frequent, just-in-time delivery of small items over short distances from supplier to user, with direct return of the container in a "closed-loop."

roughly, to evaluate how well the containers would perform in a factory environment. Buchholz facilitated the discussions that led to container selection. Here are the factors they considered:

- Accessibility of parts from assembly
- Handle/grip design: width, depth, contour
- Handle/grip location: load balance, easy access
- Label attachment method
- Ease of cleaning: drain holes, smooth surface
- Construction factors: wall thickness, durability
- Cover details: hinged, separate, king finger closure, snap closure
- Wall configuration: nestable (interlocking when stacked), straight wall

To participants in the selection process, the most important container quality was durability, Buchholz said. Containers also had to meet standards of the Automotive Industry Action Group, and they had to be made of recyclable material.

## Logistics

For the next step, logistic evaluations, Deere retained the services of Caliber Logistics (formerly Roadway Logistics, now a division of FedEx). The principal questions to be resolved at this point were

- How would parts arrive?
- How would containers be returned?
- What would return shipping costs be?
- How much of an inventory of containers would Deere have to acquire to initiate the system?

When the logistical process was well understood and cost components identified, the Corporate Returnable Container Team presented its findings and recommendations to top management, who accepted the report and approved proceeding with the project. It was August 1992. By the end of the year, after a very successful demonstration of the use of returnable containers on one of eight major assembly lines at the Horicon plant, approval was granted to expand the project to the entire plant. Now, the process of bringing suppliers into the loop began. Deere had decided to purchase and provide the new returnable containers itself rather than requiring suppliers to provide them. Buchholz explained why:

1. By removing containers from suppliers' costs, Deere would realize an immediate piece price reduction in manufacturing costs.
2. Deere could decide how many different containers it wanted. (The final answer was nine.) If suppliers had made individual decisions, the number of variations would have been much higher.
3. Staging the return of containers would be more efficient by drawing them from the Deere pool of standard sizes rather than segregating and returning containers owned by individual suppliers.

Suppliers were extremely receptive to the new program, Buchholz reported. The returnables program meant that suppliers no longer had to purchase and inventory corrugated cardboard cartons, pallets, dunnage, etc. Gone was all the labor associated with setting up various kinds of transport packaging. Stretch wrapping and banding of loads became easier with the more durable containers.

## Costs and Benefits

By 1997, Deere had invested about $8 million in the returnable container program at the Horicon plant. Annual costs are about $300,000 a year—for new containers, new parts, increased volumes. But the program also produces substantial annual savings. In 1997, it was anticipated that through 2000 (a year selected only because it marked the end of the 10-year project period) the use of returnables would reduce production costs by about $1.9 million a year. In terms of payback rate, when the project began in 1992 the payback period was calculated at 2.7 years, mainly through reductions in piece prices. In fact, cost recovery took a little less time, 2.5 years. Stated otherwise, Deere's return on investment was 71%.

The program began with an initial purchase of about 12,000 containers. Since then the inventory has expanded to 110,000 containers in nine standard sizes, the smallest measuring 12 × 7 × 5 inches, and the largest with a pallet footprint,

48 × 45 inches, with collapsible side walls 35 inches high. As to direct benefits along the assembly lines, Buchholz lists the following:

- *Greater efficiency and improved ergonomics* by delivering parts close enough to the production line that employees can work directly out of the shipping container.
- *Safer operations.* Since it is no longer necessary to cut corrugated cartons open with utility blades, the number of accidental lacerations has fallen significantly.
- *Improved part protection* compared to shipment in expendable cartons.
- *Improved housekeeping*—less cardboard dust, fewer wood chips from crates.

## Managing the Container Inventory

The 110,000 returnable–reusable parts containers represent only a portion of Deere's long-term program to eliminate expendable containers. Since 1994, the company has shipped finished tractors to dealers in Envirocrates, Deere's own design of a reusable shipping crate, and manufactured by Deere at Horicon. The Envirocrate eliminates crating made of corrugated fiberboard and wood. Including Envirocrates, Deere now has about 296,000 reusable containers in more or less constant motion between its suppliers, the Horicon plant, and dealers. To manage container flow, monitor container inventory and reorder as necessary, and handle maintenance and repair, Deere selected Caliber Logistics as its contractor. In fact, Caliber played an important role in designing and implementing the entire returnable container program. The logistics company has established container flow through the Horicon plant, including the design and maintenance of a tracking–restocking program that connects each assembly line station to supplier locations and all intermediate handling or transportation points. Of key significance, the system enables Deere employees to order daily quantities of parts direct from their work stations simply by scanning the part's bar code I.D. with a laser. No other employees are directly involved in the restocking process. The system has enabled Deere to reduce inventory levels by 18%, a significant portion of overall cost savings.

## Doing the Right Thing

Obviously, John Deere & Co. takes great satisfaction in the reduced costs and improved efficiency that have stemmed from the returnable container program. Equally, the company is proud as a corporate citizen to meet the goals of the U.S. EPA 33/50 Program, in which companies voluntarily pledge to reduce dependence on landfilling by at least 50%. The Horicon program has permanently removed 3,000 tons of wood and 2,000 tons of corrugated cardboard from the waste stream, illustrating how companies do good while doing well.

## GROCERY INDUSTRY EXAMPLES

In the last chapter we saw how Kroger cuts its pallet costs by contracting with third-party suppliers and by replacing wooden pallets with plastic pallets for shipments from Kroger distribution centers to supermarkets. Returnable containers also are commonly used in the grocery and supermarket industry where perishable products move from the producer direct to the store without passing through a distribution center. Milk and bread are prime examples. Returnable–reusable milk and bread cases have been used for decades. Early models in the dairy industry were made of wood; now most are plastic. Baked goods are delivered in reusable fiberboard and plastic trays. The soft-drink and beer industries offer other examples, going back many years to the wooden crates or fiberboard cases used to deliver full bottles and recover the empties for refilling. When the industry initially moved away from refillable bottles, reusable containers disappeared. But they made a comeback in the 1980s with the introduction of the 2-liter plastic bottle. Pepsi-Cola, for example, delivers a vast majority of the large drink containers in plastic crates. Cost information provided by Pepsi-Cola indicates that the purchase price of a one-trip, one-use fiberboard carton is 17% the cost of a plastic crate. But when the container cost is calculated per use, the reusable plastic crate, capable of 100 round-trips on average, costs 62% less than the fiberboard carton.

## OBSTACLES AND SOLUTIONS

Adopting returnable–reusable transport packaging is not without problems. But neither is there a lack of solutions. David Saphire pairs off the obstacles and solutions as shown in the following table.

**Expanding Reuse of Shipping Containers: Obstacles and Solutions**

| Obstacle | Solution:<br>Industry Options | Solution:<br>Government Options |
|---|---|---|
| Large initial capital expense | Third-party leasing | Low-interest loans, tax credits |
| Cost of administering system | Third-party leasing<br>Industry-wide standardization of containers | Mandated standardization |
| Cost of long-distance freight to return containers | Industry-wide standardization of containers | |
| Resistance to changing packaging and shipping practices | Cooperative effort within an industry to work with suppliers | Government procurement guidelines that favor or require reusable shipping containers |
| Lack of procurement clout on the part of small business | Cooperative purchasing | |
| Lack of storage space for empty containers | Collapsible, nestable, and stackable containers<br>Frequent collection of containers<br>Just-in-time delivery<br>Direct delivery | Financial incentives for establishing storage depots, such as loans or tax credits |

*Source:* David Saphire, *Delivering the Goods: Benefits of Reusable Shipping Containers,* Inform, Inc., New York, 1995

## CASE STUDY: Corning Inc.

Corning, New York

*Returnable pallet system eliminates 233 tons of corrugated containers, reduces packaging material storage requirements by 35%, cuts handling time, reaches break-even after two roundtrips.*

Corning Inc. is a major manufacturer of catalytic converters for diesel truck and car engines. There are three basic steps in the manufacturing process. First, Corning produces a ceramic substrate; then it ships the substrate to a catalyzing company for coating; and finally the coated substrate is sent to a "canning" company, where the substrate is encased in a stainless steel jacket, readying the product for installation in the exhaust system of a car or truck.

Before adopting a returnable container system, components were shipped from point to point in single-use corrugated cardboard cartons. Besides consuming a large amount of material—233 tons of corrugated disposed each year at the canning point—the old process also consumed time. Corning required one person per line just to erect corrugated cases. At the canning site, Delphi E, in Flint, Michigan, time was required to unload converters from cases, dispose of empty cases, and control corrugated dust. Housekeeping alone was calculated at 650 hours per year. Removing the substrate from the carton and disposing of the carton took about 15 seconds per converter, or about 1-1/2 hours per day for each person handling substrates.

The new transport packaging system is based on a plastic pallet measuring 45 × 48 inches. Thermoformed from a single sheet of 3 mil polyethylene, the pallet has nine legs, allowing four-way entry, and a load capacity of 3,000 pounds. The pallet also can be turned upside down and placed on top of the load as a cap; and the staggered layout of the legs permits the pallet-cap of one load to interlock with the pallet-bottom of another. Loads can be stacked four-high.

Catalytic converter substrates are loaded in layers at Corning, with a die-cut foam pad between layers and a 1/4-mil HDPE tray, embossed with cavities to hold individual substrates. The tray also can be inverted as a cap, and it grips the foam pad cushion through a pattern of "buttons" in the tray that nest in a pattern of die-cut voids in the pad. This permits loads to be moved around the Corning plant without stretch-wrapping.

Previously, four catalytic converter units were packed per corrugated carton; each carton weighed 35 to 40 pounds—a heavy load for line workers to hoist multiple times per day. Now, with units packed in open layers on a pallet, only one substrate has to be handled at a time, an ergonomic improvement. The elimination of corrugated cases and inner dividers has freed up 35% of warehouse space at Corning previously dedicated to this one product. The reusable packaging components are returned from Delphi to Corning by common carrier. Empty pallets, trays, and pads occupy about one quarter of the space compared to full loads of product.

*Payback*

On a per-unit-of-product basis, Corning's reusable shipping package costs about twice as much as the single-use corrugated boxes they replaced. Filling the transport pipeline with the new units required a large up-front investment. However, the cost of each new shipping unit was recovered in only two round trips. Normal wear and tear requires annual replacement of 10% of containers. Otherwise, the reusable packaging units are expected to last indefinitely.

## CASE STUDY: Chrysler Corp.

Belvidere, Illinois

*Plastic closures installed to prevent contaminants from entering auto components are recovered and reused, saving the company $130,000 a year in delivery and disposal costs.*

Chrysler Corp. has been implementing waste prevention programs since 1989, with a goal of reducing landfilling to zero by 2000. The Belvidere plant employs 3,700 assembly-line workers to produce the Dodge Neon. When the waste stream was examined for opportunities to reduce cost, it was observed that the numerous protective closures inserted into components arriving from outside the plant possibly could be reused instead of discarded. To set up such a program the company had to persuade suppliers that (1) the closures (made of LDPE—low-density polyethylene) really were reusable and (2) they could be returned in sufficient quantity and at the right time to enable reuse.

Collection of plastic closures occurs along the assembly line. They are removed and tossed into Gaylord-style cardboard boxes measuring 48 × 45 inches. Full boxes are removed to an area near the loading docks where closures are inspected for contamination and damage, and sorted by color. Seven vendors participate in the reuse program. Typically, they collect their closures once a month, back-hauling on the same trucks used for deliveries to Chrysler; thus, transportation costs are avoided. Experience indicates that closures can be reused up to 10 times.

*Payback*

Chrysler reports it recovered the cost of establishing the reuse program in 6 months, by (1) reducing disposal costs and (2) receiving a credit against vendor freight charges for the value of plastic closures reused. The following table illustrates the second point.

*Future Consideration*

Chrysler and its vendors are discussing reducing the number of different colors of closures where there is no particular reason to distinguish by color. Already the

| Material Returned | Cost Per Item | Number Returned Per Car | Yearly Savings |
|---|---|---|---|
| Oxygen sensor (tray) | $0.35 | N.A. | $ 4,500.00 |
| Gas tank cap | $0.02 | 1 | $ 5,336.12 |
| Crank shaft cover | $0.11 | 1 | $ 29,348.66 |
| Throttle body cover | $0.07 | 1 | $ 18,676.42 |
| Input shaft cover | $0.19 | 1 | $ 50,693.14 |
| A/C covers | $0.02 | 1 | $ 5,336.12 |
|  |  | TOTAL | $ 113,890.46 |

SAVINGS PER VEHICLE BUILT: $0.4269

plant has realized additional savings of $21,000 by switching to clear plastic for some closures.

### CASE STUDY: Prima Frutta Packing Co.

Linden, California

*Wooden apple crates used to move fruit in bulk from the orchard to the packinghouse are replaced with high-density polyethylene (HDPE) bins, reducing crate maintenance and replacement costs, increasing warehouse capacity, and reducing damage to apples.*

Prima Frutta Packing Co. grows and packs produce including apples, pears, cherries, and walnuts. Most of the apples are shipped to U.S. and Canadian markets; the remainder to overseas markets. Providing perfect apples, such as Granny Smith, Royal Gala, and Fuji varieties, is a matter of pride for Rich Sambado and his brother Tim, who operate the family-owned business.

Previously, apples picked in the company's Northern California orchards were placed in 4 × 4 × 2-foot wooden crates for bulk transport a short distance to the packing facility and warehouse. After a bumper crop of apples a few years ago required the purchase of hundreds of new crates, Prima Frutta decided to switch to a new crate that would last longer and cool the fruit better. Following experimentation and testing, the company settled on injection-molded HDPE bins. Prima Frutta has converted 100% to the HDPE bins for the apple crop. The large white containers hold 24 cases of fruit weighing 40–50 pounds each.

### Payback

Although HDPE bins of the same size cost $80 each compared to $60 for wooden crates, Prima Frutta anticipates long-term savings. The wooden crates required frequent repairs, deteriorated from exposure to the weather, and were more difficult to clean. In addition, with the price of wood rising, the difference between wood and plastic had been narrowing. After 3 years of experience producing HDPE crates, Macro Plastics says its bins will withstand 15 to 30 years of service, depending on how they are handled.

The HDPE bins are stronger than wooden crates and are designed with rounded inside corners, providing better protection of fruit. This reduces the proportion of bruised fruit—"juicers"—and increases the proportion of apples acceptable for shipment.

### Additional Waste Prevented

The HDPE bins have a rigid corner column design allowing these crates to be stacked twice as high as the wooden crates they replaced. The new bins are strong enough to withstand nearly 5 tons of compression weight, sufficient for an 8-bin stack. They also are designed to interlock, making them less likely to topple over. These design features have increased cold-storage capacity by 30–50%. In addition, built-in forklift guides help avoid damage to the base of the plastic bin. The new bins are easier to sanitize than the old wooden crates, which could absorb chemicals used in the orchard. Prima Frutta uses the apple bins for pears and is working with a supplier to develop additional special bins for bulk handling of other fruit.

### CASE STUDY: Cook, Inc.

Ellettsville, Indiana

*A supplier to the health industry replaces single-use corrugated cardboard shipping containers with reusable, multipurpose plastic tubs for three-point movement of materials, cutting annual costs by $6,500.*

Cook, Inc. manufactures standard and special-order medical instruments for a world market. Some instruments contain electrical components requiring assembly of wires, connectors, and guides. This work is performed by a local nonprofit organization, Stonebelt, Inc., an employer of persons with physical and mental handicaps.

Cook's logistical system begins with the shipment of raw materials—wire, connectors, holders, plastic parts—from a vendor in Illinois and from Sabin, Inc., a Cook subsidiary, to the main Cook plant, where parts required for assembly at Stonebelt are reloaded for transhipment to the assembly contractor. After assembly, the medical instruments are returned to the Cook plant for warehousing and order fulfillment.

Originally, parts and products all moved from point to point in corrugated cardboard cartons that were used once and discarded. That container system was replaced by the use of plastic tubs, which circulate continuously between parts suppliers, assembler, and Cook warehouse. It was found that an inventory of 30 reusable tubs was sufficient to replace an average monthly consumption of 153 single-use corrugated cartons.

### Payback

Cook is able to use one size of tub for all transfers of parts and products. Each tub costs slightly more than $100. Thus, a $3,000 one-time investment in reusable tubs (30 × $100) compares to an annual expense of $6,500 for expendable cardboard cartons. The company recovered its investment in about 6 months.

## Additional Benefits

1. Since Cook supplies the required shipping containers, Cook's vendors no longer have to be concerned with purchasing them—reduced paperwork for purchasing and inventory.
2. Cleaner working environment: Cook has found that eliminating corrugated cardboard cartons in these manufacturing operations has significantly reduced dust levels in the plant. This is such an important benefit that the company is working with other vendors to reduce the use of cardboard cartons.

## CASE STUDY: Delco Electronics Corp.

Kokomo, Indiana

*The Packaging Systems Group recovered and reused 30% of the incoming plastic reels of auto wiring components, reducing the cost of new reels by $150,780.*

Delco manufactures automotive electronics for engine control, anti-lock braking, air bag deployment, instrumentation and display, comfort control, suspension control, navigation and object detection, and audio systems. The company employs about 10,500 at the Kokomo plant.

In 1991, the company determined to reduce the volume of materials sent to landfill disposal. The Kokomo Operations Steering Team—an open committee of approximately 25 hourly and salaried workers—was formed and began to examine the company's waste stream to see what might be recovered for recycling or, better still, eliminated by source reduction. Later the team also found markets for recovered materials, organized collection and transportation systems, and educated plant workers in waste reduction programs.

Waste stream analysis revealed a large volume of component reels being disposed. These are plastic reels resembling motion picture reels but used to hold continuous strings of electronic components. Along the assembly line, reels feed machines that automatically place electrical wiring in cars. Reels vary up to 13 inches in diameter and up to 18 inches wide. Many are made of polystyrene (PS, No. 6), some of combinations of plastic and paperboard. Delco was accumulating empty reels at the rate of 2,000 a day.

Delco set an objective of returning the maximum number of reels to suppliers for reuse and reusing reels internally. One of the company's suppliers agreed to reuse reels (but the others did not, citing concerns over quality, contamination, and static electricity). The reuse process began with collection of empty reels in a gaylord box at the assembly line and then delivery to a refurbishing contractor—a sheltered workshop 1 mile away from the Delco plant—for sorting by size and color, cleaning, removal of labels, spraying with anti-static chemical, inspection, and just-in-time shipment to the supplier for reuse. Internally, reels are reused for interdepartmental transfer of components, averaging 10 reuses.

Reuse by the outside supplier proceeded successfully for a year but was discontinued for lack of a sufficient supply of reels to meet JIT demands, and because of transportation costs to the supplier's location in Texas.

### Payback

An estimated 30% of incoming reels are reused. The cost to refurbish a reel for reuse is 58 cents; the cost of a new reel is $7.50. During 1995, Delco sent 22,656 reels for refurbishing and realized net savings in new reel purchases of $156,780. Reels that are not reused internally are sent to a plastics processor for grinding and recycling. During 1995, Delco reused or recycled 280 tons of component reels.

In mid-1996, the company purchased a new labeling machine, enabling the use of peel-off labels on reels. This is expected to speed the refurbishing process.

### CASE STUDY: Ben & Jerry's Homemade, Inc.

## South Burlington, Vermont

*After studying the alternatives for delivery of certain bulk ingredients, ice cream factory selects a reusable 275-gallon container, thus increasing product per pallet, reducing warehouse space requirement, and avoiding the disposal of 800–900 pounds per day of corrugated cartons.*

As Ben & Jerry's Homemade, Inc. prepared to roll out a new flavor—Holy Cannoli®, a vanilla ice cream with an Italian accent based on ricotta cheese, roasted pistachios, and cocoa-coated cannoli pieces—one logistics question concerned how to deliver the ricotta in bulk, from a source 1,500 miles away. The main choices were as follows:

- 5-gallon bags-in-boxes (BIBs). If this were the choice, the company would need to receive 90 such BIBs daily and manage the recycling or disposal of the containers, including 800 to 900 pounds per day of corrugated cardboard. The small containers required a lot of in-plant handling, and at 35 pounds each, they seemed to come with back strain built in.
- 55-gallon drums. As a trial run revealed, workers often had to manually saw the metal rims from drums before they could be disposed—a high potential for accidents.
- A returnable–reusable, 275-gallon bag in tote. This steel-framed intermediate bulk container (IBC) was the ultimate choice, for a number of solid reasons:

  - Service life of up to 10 years.
  - Compared to BIBs, increases product per pallet by at least 50%.
  - Strong enough to be stacked five high.
  - Collapsible to about one-quarter total size for return shipment.
  - Ergonomically correct, saves time and labor.
  - Leaves nothing behind, nothing to dispose of.
  - Equipped with its own galvanized steel, 48 × 40-inch, four-way pallet.

- Equipped with a 3-inch valve, allowing virtually all of the viscous ricotta cheese base to vacate the bag.
- No capital investment in containers.

## Payback

Ben & Jerry's employees have been spared a great deal of repetitive motion associated with using smaller containers. But the biggest benefit of using the 275-gallon IBC is avoiding 900 pounds of paperboard packaging and 360 plastic liners each day simply to produce one flavor of ice cream.

## CASE STUDY: General Electric Corp.

### Bloomington, Indiana

*By substituting reusable steel frames for bulk shipment of glass shelves for refrigerators, 157 tons per year of wooden frames and cardboard boxes are eliminated, reducing damage losses, handling, and disposal expense by about $50,000 a year.*

GE assembles and ships nearly 940,000 refrigerators per year at its Bloomington plant. The company was experiencing an unacceptable level of damage to the glass it received in bulk from a supplier in Vincennes, Indiana, approximately 72 miles distant. Analysis showed that damage occurred primarily due to the use of wooden crates and wooden dunnage, which permitted some shifting of the load during truck transport.

Discussions with the glass supplier, Gentron Corp., revealed the vendor's similar interest in reducing costs and losses. Thus a search was initiated for a suitable new crating system. Gentron agreed to cover the up-front cost and to recover the cost later as part of its pricing structure to GE. A reusable steel frame was designed and tested and put into service. It measures 52 × 45 × 51 inches and is sized for direct palletization. Each frame has a capacity of 1,000 pieces of refrigerator shelf glass. Empty frames deadhead back to Gentron on the same truck used for delivery.

This change in bulk shipping container has saved GE more than $50,000 in combined savings from reuse of the bulk container, reduction in product damage, and elimination of disposal of single-use wooden crates and corrugated cardboard and the associated labor. Waste disposal tonnage dropped by 157 tons.

Another change in shipping routine is under evaluation. It would entail replacing, with plastic shrink wrap, the wooden lath and steel strapping presently used to secure glass plates to the steel frame, further reducing the volume of waste. In addition, the full sheets of butcher paper presently used to separate individual sheets of glass would be replaced by 1-inch diameter, stick-on tabs.

## Payback

The one-time cost of steel frame construction will equal the cost of equivalent single-use wooden crates and dunnage in about 12 months.

**Figure 2**   $50,000 ADVANTAGE. A steel frame replaces throw-away crates to transport glass shelves for refrigerators. GE's waste disposal tonnage dropped by 157 tons a year. Product damage also declined.

## Other Benefits

1. Material disposal labor time has been reduced.
2. Use of shrink wrap will permit easier rewrapping of a load of delivered glass, compared to steel strapping, after a partial amount of glass has been removed from the steel transport frame.
3. Use of shrink wrap also will eliminate cuts on workers' hands from sharp edges of steel bands flying loose upon opening a crate.

# CHAPTER 6

# Teaming with Suppliers and Customers to Prevent Waste

Without ever looking outside the organization for help, companies can achieve significant reductions in waste. But they can achieve even more by enlisting suppliers, customers, and community organizations in a common effort. That's the theme of five case studies presented in this chapter.

It may take a leap of faith for companies to seek outside partners in waste prevention. Depending on the company culture, there may be either a willingness to venture beyond the security of home territory, or great reluctance to do so. The very idea of doing so might be dismissed as too risky. It's easy to determine a company's inclinations in this regard: look at the way people work with one another within the organization. Rockwell-Collins Avionics & Communications illustrates the point. This electronics manufacturer in Iowa felt confident about forming an internal waste prevention team with key members from two of the most sensitive—and in some company settings, the most polarized—company groups: organized labor and purchasing. This made perfectly good sense. Who knew better where waste was occurring than people on the production line? And who was in a better position to do something about waste related to the ordering and delivery of raw materials than people in the company purchasing function? This high level of internal trust made it comfortable for Rockwell-Collins to bring Goodwill Industries into the loop to act as facilitator of a cooperative waste prevention program between the company and many of its suppliers.

EMC Corp. and Bayer Corp. provide good models for the recovery of packaging for reuse. However, their objectives were quite different. EMC saw an opportunity to reduce the cost of a fairly expensive shipping container, while Bayer was motivated more by environmental concerns, both its own and its customers'. Bayer simply wanted to break even (and it came close).

StorageTek-Network Systems Group and NACHI Technology, Inc., illustrate the triumph of common sense over trash. In each case, the company took a fresh look at the way materials were arriving from suppliers and immediately saw opportunities to reduce costs by adopting "pass-through packaging"—the same carton used to deliver components to the assembly line also transports the finished product to the buyer.

This discussion may seem to imply that most good ideas about waste prevention come from the enterprise in the center, the manufacturer. Not so. Suppliers and customers both have much to contribute. But all enterprises are cast in various roles depending upon the point of view. NACHI Technology, for example, considers General Motors a customer; but viewed from GM, NACHI is a supplier, and NACHI understands being in this other position very well, illustrated by the way it has enhanced its position with GM by finding innovative ways to cut costs—in part by going to *its* suppliers for help. If cooperative relationships do not occur spontaneously, perhaps they can be primed. In Chapter 3 we saw how Columbia University uses a letter-survey to invite vendors to participate. Here is an excerpt from another example, written by the U.S. EPA WasteWi$e office and based on an actual letter sent by a large retailer to its suppliers.

Dear Valued Supplier:

Our company and its suppliers enjoy a proud tradition of anticipating and responding to the challenge of evolving customer demands. Today, that challenge is even more complex due to increased competition, cost-cutting measures, and the emergence of a new force—environmental issues. We share the national concern for improvement of the environment and preservation of our natural resources. We are committed to reducing our company's impact on the environment and satisfying customer demand for environmentally sensitive products and packaging. This commitment will impact virtually everything we sell and service.

Recent surveys indicate that customers increasingly make buying decisions based upon environmental factors, and product packaging is a large component of our country's solid waste. Reducing, reusing, and recycling packaging and other materials can significantly cut purchasing, operating, and disposal costs.

Specifically, we ask you to be a partner with us to:

- Look for opportunities to reduce packaging volume and weight by reducing packaging materials used on the products you manufacture by at least 10 percent over the next two years.
- Investigate opportunities to utilize reusable shipping materials, including durable pallets and reusable plastic crates, totes, and corrugated boxes....
- Improve the recyclability of non-reusable shipping materials.

Please do not limit your efforts to these goals—be innovative! These goals should apply to all product packaging as well as any repair and replacement parts packaging you supply to us.

The above letter concerns both transport packaging and retail packaging. In Chapter 8 we'll see the ultimate retail container, after all the wraps are off.

### CASE STUDY: *Rockwell-Collins Avionics & Communications*

Cedar Rapids, Iowa

*Suppliers' cooperation helps company reduce solid waste disposal from 700 tons per year in 1992 to less than 300 tons in 1996, even as sales volume increases 25%.*

Rockwell-Collins manufactures advanced avionics and airborne/mobile communications systems for commercial and military applications worldwide. It receives a wide variety of electronics and other assembly parts in specialized plastic, metal, and cardboard containers. Before 1993, much of this packaging material was considered trash at the receiving end. Then a major new emphasis was placed on waste prevention, including an item-by-item review of packaging to determine its potential for reuse. This cooperative venture included the International Brotherhood of Electrical Workers (IBEW), the Purchasing Department, and suppliers. Implementation of the resulting program was entrusted to employee teams identified as "Solid Waste Environmental Leadership and Learning," or SWELL. In addition, a partnership was formed with Goodwill Industries of Southeast Iowa for the employment of handicapped persons to sort and refurbish parts containers.

Typical of results is the system to reuse small cardboard boxes and plastic bags (on the order of 3 × 4 × 8 inches and 6 × 8 × 12 inches). Empty containers are collected commingled and placed in gaylord boxes near the loading dock. The Goodwill truck comes by weekly and collects a dozen or more boxloads for sorting, inspection, and preparation for return to suppliers. Inspection includes, for example, testing of anti-static bags to make sure they remain reliable barriers to static electricity. The project to reuse small boxes alone has reduced disposal by 6.5 tons per year.

Other materials returned to parts suppliers for reuse include several sizes of electronic parts trays, anti-static foam, and plastic "bubble-pack" packaging. Items that cannot be returned to suppliers, including various containers, bottles, boxes, reels, and tubes, are made available free to non-profit groups and schools.

Another basic change involved the Purchasing Department. Previous policy emphasized the unit discounts available for purchase of large quantities. But the SWELL team observed that this practice also led to considerable amounts of unused and waste material, ultimately requiring disposal. The new approach permits buyers to purchase only the amount required, at slightly higher unit prices, for specialized uses when warranted. Also, suppliers have been encouraged to stop using glue to bond polyfoam to shipping boxes, enhancing the possibilities for reuse of materials.

### Payback

Putting all waste prevention programs in place cost about $60,000 in capital equipment. Rockwell-Collins pays Goodwill about $50,000 a year for its services. But measured against an annual reduction in disposal of nearly 450 tons, Rockwell-Collins considers the cost minimal. The company has prepared a document titled

"World Class Resource Recovery & Recycling Program Guide." It is available through the U.S. Navy's Best Manufacturing Practices site www.bmpcoe.org.

## CASE STUDY: EMC Corp.

### Hopkinton, Massachusetts

*A prepaid take-back kit enables customers to send packaging components back to the manufacturer for reuse, reducing packaging costs 20%.*

EMC Corp. designs information storage and retrieval systems for mainframe and midrange computing environments. The company employs 2,000 people at its headquarters in Massachusetts.

The genesis of the take-back program occurred several years ago when packaging issues—in particular, the reduction of packaging waste—were under discussion. EMC soon developed the "We Care Kit" return program for packaging of small- to medium-sized equipment weighing 90 to 115 pounds. From inception to implementation, the program took 5 months.

The program operates through the Shipping Department. Shipments to customers include a return kit, including a diagram showing which packaging materials can be returned (plastic skids, molded foam plastic, and some of the corrugated) and how to assemble these components for return. Also in the kit are two heavy rubber bands to secure the package; a self-addressed label to EMC's packaging vendor, Tuscarora Plastics; instructions for filling out the shipping form; and an 800 number to call for pick-up. The package weighs about 24 pounds, light enough for most customers to handle without difficulty. Outbound freight and return packages both are handled by an overnight delivery company.

Upon return to Tuscarora, EMC's packaging is checked to make sure it is complete and suitable for reuse. Then it is routed to EMC's Shipping Department. Materials unsuitable for reuse are recycled by Tuscarora.

### Payback

Concerning all EMC shipments that include the "We Care Kit," 35–40% of packaging is returned for reuse. Tuscarora covers the cost of inspecting returned packaging. The plastic skids, which cost $16 new, are readily reused; molded foam cushioning can be reused two to three times; corrugated cardboard has the shortest lifespan. EMC calculates the program, overall, has reduced its cost of new packaging by 20%.

### Additional Benefits

1. Customers can avoid the time and expense of recycling or disposing of EMC packaging by sending it back to the manufacturer—at no cost.
2. EMC's Purchasing Department, which tracks the program through its records of packaging purchases, appreciates the reduction in paperwork that accompanies reduced numbers of purchase orders.

3. Success of the "We Care Kit" with small- and medium-sized products has prompted EMC to consider adapting it to the shipment of larger products.

## CASE STUDY: Bayer Corp.

Mishawaka, Indiana

*By making it easy for customers to return an expanded polystyrene (EPS) shipping container for reuse, the company diverts a large volume of EPS from disposal, at minimal cost.*

Bayer Corp. ships certain perishable products to customers in an expanded polystyrene container with inside dimensions of $10 \times 9 \times 12$ inches (0.625 cubic feet). The container comes with a molded EPS lid with the Bayer logo embossed, and three reusable ice packs. Walls of the EPS container are 1-1/2 inches thick, and outside dimensions are $13 \times 12 \times 15$ inches. Thus, total volume of EPS is approximately 0.922 cubic feet. The EPS container fits inside a single-wall corrugated carton designed to withstand an edge crush test (ECT) of 32 pounds per square inch.

Bayer makes approximately 14,000 shipments a year in the kind of container described above. Before 1994, the company did not have a recovery program for these containers. Out of concern for the environment, including the volume of EPS

**Figure 1**   ROUND-TRIP TICKET. Envelope marked "Recycling Kit" goes out with each expanded polystyrene shipper from Bayer. It contains a prepaid return label. The container can be reused two or three times.

disposed of in landfills, the company formed a task force to design a system for recovering and reusing the polyfoam shippers. The five-member task force included representatives of the purchasing, warehousing/logistics, office services, traffic, and safety/training departments. They set a goal of breaking even while reducing the disposal rate of EPS containers.

The solution was to develop a reuse "kit"—actually, a No. 10 envelope imprinted "Recycling Kit—Please Open"—which is placed atop contents of the EPS container, just under the lid. The envelop contains a brief letter to the customer explaining Bayer's concern for the environment and how the return program works. The envelope also contains a United Parcel Service "UPS Authorized Return Service" label, pre-addressed to the Bayer distribution center. The label is printed by UPS and furnished to Bayer at no charge. When the EPS container is emptied of contents, the customer simply closes and seals it and affixes the pressure-sensitive label. UPS delivers the empty back to Bayer at Bayer's expense. The letter to customers suggests they reuse packaging "peanuts" (95% corn starch) if any are included; reuse the ice packs or return them to Bayer; and recycle (not return) the EPS container if it has become contaminated during shipment.

### Payback

Bayer inaugurated the new program in October of 1994. Through April 1995, overall returns to five company distribution centers equalled 28.9% of shipments—about what the task force had originally projected. Few returned containers were unusable; most could be used for three round-trips before they showed wear and were recycled. Since there was no change in the container, simply the establishment of a system of retrieval, the out-of-pocket cost to Bayer was minimal. In terms of return freight cost versus purchase of a new container, Bayer came close to its objective of breaking even. The UPS charge for return was $3.50; cost of new container, $3.13. Thus, the recovery program cost Bayer 37 cents per unit returned. The company considered this cost justified as an environmental initiative and believes it also benefited from improved customer goodwill.

### CASE STUDY: StorageTek-Network Systems Group

Brooklyn Park, Minnesota

*With the participation of a supplier, the company replaces three disposable corrugated cartons with one reusable carton, saving a total $8,450 per year.*

StorageTek-Network Systems Group (NSG) employs 600 people in the manufacture of computer hardware for sophisticated applications by business establishments. Like many enterprises, NSG connects three main locations into one orderly production process:

1. Suppliers' production sites (numerous, but we'll use just one to illustrate).
2. The NSG plant, where suppliers' parts are assembled into finished products.

3. Distribution centers, where bulk shipments from NSG are sorted for delivery to customers.

One NSG product, a piece of networking apparatus, is designed to mount in a rack. The chassis that contains the product and provides a means for rack-mounting is produced by a supplier. Previously, the supplier packed the chassis (volume, about 3.5 cubic feet) in a corrugated box for shipment to NSG, one chassis per box. At NSG, a dock worker removed the chassis, disposed of the shipping carton, inspected the part, and wheeled the unpackaged unit to the assembly line. Though not exceptionally heavy, the chassis is weighty enough when lifted repeatedly from a carton. After assembly was finished, the product was packed in a heavier corrugated carton for shipment to the customer. In some cases, this second carton was replaced by still another carton at the distribution center.

With agreement of the chassis supplier, NSG revised the process. Now, the "customer" carton is delivered to the supplier to use for transporting chassis to NSG. The inspection at the dock is eliminated—the chassis is delivered direct to the assembly line in the carton. And since the assemblers have access to vacuum lifting devices, removing the chassis causes no back strain. When the product is completely assembled it is placed back in the original carton for delivery to the customer.

## Payback

By replacing three cartons with one, NSG reduces container costs $4,000 a year. In addition, the company calculates savings of $3,500 a year in reduced time required for ordering, warehousing, and delivering containers to the assembly line, and $700 more in time saved at assembly. Finally, the company cuts its waste hauling expense by an estimated $250. Grand total, $8,450.

## CASE STUDY: NACHI Technology, Inc.

Greenwood, Indiana

*A supplier of ball bearings to the auto industry reduces transport packaging costs $25,600 a year—and disposal costs $6,800—by persuading suppliers and customers to adopt common shipping containers.*

NACHI Technology, Inc. manufactures approximately 60% of the air conditioning compressor bearings used in passenger and light trucks in the U.S. The company employs 80 people and has sales of about $20 million a year. As a key supplier to the automotive industry, the company is acutely aware of the need to control costs. Recently it focused upon packaging as a possible way to reduce materials and disposal costs and improve profit margin.

Originally, NACHI's customer General Motors required both a special pallet size (32-3/4 × 40-3/4 inches) and a special corrugated box size (19-1/2 × 12-1/2 × 6 inches) for the delivery of bearings. After discussions with raw material suppliers and the customer, NACHI found that a pallet of more standard size—39 × 45 inches—and a

**Figure 2**   3-BASE HIT. Bearing manufacturer persuades both vendors and customers to use the same corrugated carton. Thus, one box moves from raw material source through manufacture to customer. $32,400 saved.

"NACHI Box" measuring 14-3/4 × 9-1/2 × 6 inches would be acceptable for both delivery of raw material and shipment of finished bearings.

In addition, NACHI found that it could reuse the expanded polystyrene (EPS) dunnage in incoming freight as dunnage for outbound freight, eliminating the need to purchase 100,000 EPS packing blocks per year. NACHI also sought and received approval of another customer to use an incoming corrugated carton containing rubber seals as the outgoing carton for finished bearings, reducing the need for new boxes by a further 720 boxes a month.

NACHI's principal savings can be summarized as follows:

| | |
|---|---|
| Savings on new pallets for shipments to GM | $ 4,600 |
| Savings on new pallets for shipments to Ford | 5,000 |
| Savings on corrugated boxes to GM | 11,000 |
| Savings on polystyrene packaging blocks to GM | 5,000 |
| Reduction of non-hazardous compactor waste | 3,800 |
| Disposal savings from recycling old cartons | 3,000 |
| Total annual savings | $32,400 |

## Payback

Payback must be considered virtually immediate, since little investment was required outside of negotiating time with suppliers and customers. Two additional benefits should be noted:

1. Corrugated boxes used to ship finished bearings originally contained 252 units and weighed about 100 pounds, well in excess of manual lifting limits. The new bulk pack is smaller and weighs about 40 pounds, facilitating the presentation of material at assembly points. NACHI found this was a key point in winning the cooperation of GM.
2. NACHI donates serviceable used pallets to a nearby cooperative, avoiding disposal costs for the company and pallet purchase costs for the co-op.

# CHAPTER 7

# Measuring and Reporting the Results of Waste Prevention

Talking about waste, a business reporter for National Public Radio made a very keen observation. "Companies," he said, "are coming to the simple but elegant conclusion that waste and pollution represent something they paid for but can't sell—and have to pay to get rid of!" What an improvement this statement is on the old notion that waste is simply a cost of doing business. If waste is seen not as inevitable but as a conscious choice, something paid for but unsellable, it's easy to ask, "How do we quit this outrageous behavior?"

Companies "pay" for waste when they buy stuff they don't need—packaging, production feedstock, other materials. Rarely can they recover any of this cost directly. Instead, ironically, they often pay again, for disposal, or they pass the cost of waste down the line to customers. In earlier chapters we have seen many examples of the opposite behavior: intelligent action by companies to prevent waste at the source. And we have seen how rapidly an investment in this kind of activity can be paid back. But we have not examined an important motivating force, perhaps *the* force, that drives waste prevention. It is *information,* operating data as well as background data and progress reports, for both internal and external distribution. Information about the kinds of waste that are being eliminated, their significance in quantity and cost, and the resulting savings, must be communicated constantly. Following are examples of how companies measure and disseminate data on waste management and prevention.

## NorTel

Northern Telecom created a novel and very public way to talk about its accomplishments in waste prevention. The Canadian company invented an Environmental Performance Index, or EPI, to state in one simple number how the company is doing year by year in preventing waste at the source. Arthur D. Little, the Boston consulting firm, helped devise NorTel's EPI, which is designed to account for every potentially waste-producing activity within the company. There are four main categories of information:

1. Compliance, including notices of violations, spills cleaned up, and fines paid.
2. Environmental releases to air, water, and land.
3. Resource consumption, including energy, water, and *paper* (emphasis added).
4. Environmental remediation—what the company is doing to improve.

The EPI is designed to encourage progress at NorTel toward long-term environmental goals. The benchmark year is 1993; data are weighted and year-to-year fluctuations in production are accounted for by reference to the cost of sales adjusted for inflation. In addition, all EPI data are available to the general public on the World Wide Web (www.nortel.com/cool/environ/epi). All of this reporting comes at a price: the EPI manual, which is available to browsers at the NorTel web site, is a 22-page summary document and is quite sophisticated. But NorTel thinks it's important to keep all stakeholders informed. For employees and management, the EPI can work as a reward for effort, a source of pride in accomplishments, and a nudge. For outsiders, including investors, the EPI may serve as a measure of company performance, even though it comes from an interested source. Putting the EPI on the company's home page probably is a public relations plus.

## POLAROID

Polaroid Corp. measures waste by units of production. The Cambridge, Massachusetts manufacturer of photographic products inaugurated a formal waste prevention program in 1988, which has been documented by WasteCap of Massachusetts, a partner in the EPA WasteWi$e program. Each manufacturing division of Polaroid selected a product to represent its operations and use in normalizing waste byproduct production. Goals were set: for nonhazardous solid waste, a reduction of 10% a year for the first 5 years, and 7% a year for each of the following 5 years. Between 1988 and 1993, Polaroid in fact reduced solid waste per unit of production by 31%, or a little more than 6% a year.

To keep track of its progress in waste prevention, Polaroid developed a centralized database called EARS, the Environmental Accounting and Reporting System. EARS integrates data from 23 plant sites and yields a weighted index for the entire company. It also enables Polaroid to predict the waste impact of a new product. EARS also records waste prevention in support functions, specifically,

- Reduction of paper by a number of means, including greater use of electronic data, payment on receipt, and credit card purchases.
- Reduction of inventory through just-in-time ordering.
- Reduction of packaging through conversion to returnable–reusable containers for delivery of materials.

To illustrate the final item noted above, packaging reduction, Polaroid replaced a large number of single-use corrugated fiberboard boxes with reusable "totes." Used for closed-loop shipment of components from suppliers, the totes are made of virgin fiber corrugated, have no top flaps, and fold flat when empty. They are strong enough to make 20 trips, and are then recycled. During the first year of use, totes enabled Polaroid to eliminate 100 tons of corrugated from its system and save $70,000.

## PSE&G

Public Service Electric and Gas Co., the New Jersey utility, estimates that in 1990 it was spending $10 million a year on waste management services, mainly disposal. An estimate was the best the company could do because information was dispersed, there was no consistency in tracking costs, and in some cases, no tracking at all. Now all that has changed, and PSE&G stands as an example of how information empowers waste prevention. In 1994, the company created a Resource Recovery Group within its Materials Management Support Organization. Responsibility for waste management was centralized, and the emphasis shifted from waste minimization to waste prevention. By 1995, PSE&G had cut the waste management budget in half, to under $5 million; and by 1997, this budget was under $3 million.

The utility's very dramatic reduction in its waste management budget was made possible, first, by consolidating 40 separate operating-site budgets into one; second, by reducing the number of waste-hauling contractors by approximately the same ratio; third, by implementing waste accounting; and finally, by emphasizing *investment recovery* within the waste management function. During the first 9 months of 1997, investment recovery yielded $2.64 million in savings or revenue for PSE&G. The program comprises three specific activities:

1. *Redeployment of materials.* This means using surplus items at one operating site to fill the needs of another site. Office furniture, for example. Previously, when offices were refurbished, old cabinets, desks, partitions, etc., were scrapped. Now, unusable but serviceable materials are listed as available to any PSE&G department. The success of redeployment depends on close cooperation between the Resource Recovery Group and the company procurement function.
2. *Selling surplus materials.* PSE&G conducts vehicle auctions, for example, and through broker specialists has sold surplus generator parts to utilities in the Third World. Revenue from the sale of recyclables also is included in the $2.2 million recovered this way during 1997.
3. *Donations.* When PSE&G acquired new desktop computers throughout the organization, it arranged to refurbish the old units and donate them to New Jersey schools. The value of these donations during 1997 was more than $500,000.

PSE&G states its guiding principle of waste management in a few words: "Accounting for Waste as We Do for Money." The utility defines waste accounting as "a method for systematically compiling information about wastes that permits tracking of pollution prevention performance."

## TRACKING SPECIFIC ITEM COSTS

Accounting for waste prevention may present a challenge. Unlike financial accounting, for instance, waste accounting is a new idea, often seeming to lack well-understood principles and well-identified, reliable centers of information. Yet the essential information generally is available, as the following four cases illustrate. Each of these companies is a partner in the U.S. EPA WasteWi$e program, the source of the information.

*State Street Bank & Trust Co.* Like many large organizations, this Boston-based bank frequently transfers personnel and functions from one office to another, and has a fairly continuous demand for moving services in general. For interoffice relocations and moving, the bank decided to rent reusable plastic crates instead of purchasing corrugated cartons. To determine the reduction in waste for 1 year, the bank began with an estimate from its moving consultant that a corrugated box could be used an average 2.5 times. Invoices from the crate rental company indicated the bank used crates for 25,000 trips during the year. Thus, 25,000 trips in plastic crates divided by 2.5 average trips in corrugated cartons equals 10,000 corrugated cartons eliminated from disposal.

*Perkin Elmer Corp.* This Connecticut-based manufacturer of analytical and environmental systems recovers and reuses a substantial volume of transport packaging by making it convenient and cost-free for customers to participate in the return program. To measure the amount of waste prevented, the company set up a procedure to accumulate shipping invoices for all return shipments of packaging. Thus, the number of units returned and the freight charges are known. The cost of refurbishing boxes is added to freight charges and the total cost is compared to the total cost to purchase an equivalent number of new shipping packages. The return rate averages 28%. In 1 year, the company reused 62 tons of packaging components and saved $95,000. *(See also the Bayer Corp. and EMC Corp.)*

*BankAmerica.* This major world banking organization, headquartered in San Francisco, launched a three-part paper reduction program in 1994:

1. Double-sided copying of office documents and pared-down distribution lists.
2. Elimination of procedures manuals in branch offices, substituting a centralized library and telephone support.
3. Double-sided copying of customer checking account statements. Records maintained by the bank's supply warehouse made it possible to measure the reduction in paper use. To calculate the reduction in paper consumption based on double-sided copying, the bank compared shipments to the relevant company departments in 1993 and 1994. Similarly, the total number of manuals printed in the 2 years was compared. (The number of bank employees and number of customers was approximately the same from one year to the next.) Certain factors were employed, such as 5 pounds for a ream of paper. Altogether, the bank found that it saved 750,000 pounds of paper and upwards of $1 million in 1 year.

*Southern Mills.* This Atlanta-based company established a drum take-back program with its dye and chemical vendor. Since there was only one vendor involved and all drums were returned, calculating the amount of waste prevented was easy. In a recent year, for example, Southern Mills received 883 metal drums (empty weight, 40 pounds) and 334 fiber drums (empty weight, 22 pounds). Altogether, 42,668 pounds of bulk transport containers were returned for reuse. Previously, they had been landfilled.

Tracking systems apply to more than mundane things such as pallets and corrugated cartons. The Walt Disney Co. established a computerized system to track movie production sets and facilitate their reuse. During 1995, this program saved the company $528,000 in wood purchases.

### CASE STUDY: Whirlpool Corp.

Evansville, Indiana

*Measurement is a key factor for the task force assigned to reduce the scrap rate associated with internal handling of refrigerator crisper and meat pans. Total savings, $473,000 a year.*

The Whirlpool Evansville Division is a major manufacturer of domestic household refrigerators for the North American market. This project began after in-house scrap costs and assembly line rejection rates for the injection molded vegetable pan were found to be excessive. A 20-member quality team was established and eventually recommended a number of changes to a system that had been in place for more than 20 years.

Improvements affected such areas as updating quality systems, defining quality visual standards, and defining an optimal packaging scenario. Initially, an audit was completed of scrap parts from the assembly line. Analysis reflected improper quality standards, incorrect operator handling, inappropriate packaging, etc. Based on this audit, yearly total scrap costs of $207,800 were forecast if no corrective action were taken.

Pans were found to have been nested on their side, then stacked on skids to such height that the accumulated weight caused warping. Improperly sized and configured skids resulted in damage during skid movement and scratching of pan surfaces.

Improvements implemented included the use of properly sized plastic skids rather than wooden skids, and the introduction of polyethylene separator sheets and corrugated sleeves to eliminate pan scratching and warpage. Damage during skid movement was reduced 90.5%, and skid pack capacity was improved by the same percentage.

Whirlpool notes that this problem was solved by a cross-functional team. Initially, no one wanted to try a new packaging scenario. However, the team continually strived to define optimal packaging requirements of all partners in the process, earning the support and commitment of all participants.

## Payback

Whirlpool estimated the first-year benefits, all inclusive, at $473,400.

1. The components were:
   - Scrap rate reduced 64.9%, or $60,000 per year
   - Pack capacity increase, 90.5%
   - Damage in transit reduction, 90.5%
   - Packaging weight reduction, 40%
   - Increased storage capacity owing to pack height reduction, 29%

Total cost productivity improvement for the division of 5% yielded a savings of $302,401. The internal rate of return was 70.23% with a payback period of 2 years.

## Other Benefits

1. The scrap rate reduction indirectly reduces overall energy costs—fewer replacement parts have to be produced. Also, plastic feedstock costs are reduced.
2. Weight reduction of approximately 40% permits manual movement of skids, a time-saver in some situations.
3. Approximately 120 salaried and 1,500 hourly employees were trained in defining visual quality standards using actual standard samples. Copies of visual standard method sheets were provided to all areas of the assembly line for continual reference (an excellent example of effective communication).
4. For all its efforts in waste prevention, Whirlpool was selected to receive the Indiana Governor's Award.

## CASE STUDY: AAP St. Marys Corp.

St. Marys, Ohio

*Precise measurement of material and process costs confirms the wisdom of a change in production: monthly material savings, $34,860; process savings, $94,630.*

AAP transforms aluminum ingot into finished cast aluminum wheels for both domestic and foreign automotive vehicles. Machining produces a significant volume of scrap aluminum chips—about 6 pounds per wheel, or roughly 30% of the original casting. Previously, this scrap was sent to an off-site aluminum recycler where chips were melted to produce an aluminum ingot. On-site chip recycling eliminates the need to re-form ingot from chips, thus eliminating one complete melting cycle.

*Estimated energy savings.* AAP's melting process uses approximately 1,000 BTUs per pound of aluminum, and it is estimated that the outside contractor's usage would be about the same. Since AAP remelts 9,960,000 pounds of chips per year, net annual energy savings amount to 9.96 billion BTUs of fuel. With annual production of 1.5 million wheels, savings can be expressed as 6,640 BTUs of fuel per product unit.

*Estimated waste reduction.* Dross is the waste byproduct of aluminum melting operations. AAP's new remelting process reduces dross—from 8% dross under the old process to 2% dross under the new in-plant recycling process. This can be expressed as 0.40 pounds of waste dross reduction per wheel.

## Payback

*Material* cost savings from AAP's process change are significant. By remelting chips in-house, AAP increases the yield to 98% from 92%. Monthly savings can be calculated with the following formula:

Savings = Chip Volume × Yield Differential × New Ingot Cost

Savings = 830,000 lb × 6% × $0.70/lb

Savings = $34,860/month

*Process* cost savings can be expressed as follows:

Old Process Cost = Chip Volume × Yield × Outside Process Cost/Pound

Old Process Cost = 830,000 lb × 92% × $0.175/lb

Old Process Cost = $133,630

$133,630 less New Process Cost ($39,000) = $94,630 Process Cost Savings

# CHAPTER 8

# Retail Packaging: Dispelling Myths

Packaging myths flourish at the consumer level. Ask a sample of nonexpert retail shoppers about product packaging and you will hear general agreement that most items are "over-packaged." It's easy to reach such a conclusion while standing in the check-out line, holding a blister-packed kitchen utensil, for example. The package may be three or four times larger than the product—prima facie evidence of waste! What gets lost in this instant indictment is the way that mass merchandising works these days, with a minimum of human intervention and maximum of self-help. The big blister-pack provides space for instructions (and disclaimers), a deterrent to shoplifting because of its sheer size, and, perhaps most important, a hole for hanging the product on a display rack.

Other myths abound. In a study of packaging waste by the Institute for Policy Analysis of the University of Toronto, researchers found popular support for these ideas:

- *Landfill scarcity is the main reason to reduce packaging.* In fact, there's plenty of space for landfills in North America (though some sites require a trip).
- *Source reduction doesn't accomplish much compared to other packaging strategies.* Broadly speaking, untrue. In the Canadian soft drink industry, for example, source reduction leads recycling three to one in waste diversion.
- *Strong regulatory action, like deposit–refund systems, is justified to achieve packaging reduction.* In fact, simply making the container lighter is far more effective both in terms of waste-prevention and cost than setting up a deposit system.

But the biggest myth is the assumption that manufacturers deliberately "over-package." No one keeps statistics on the time and money that industry invests in designing, testing, and producing packages that consume *less* material, *less* labor, and *less* money, but the totals must be enormous—because the *payback* for preventing packaging waste is so great. Like the power of compound interest, a few pennies saved on each retail package generate huge savings when multiplied by the total number of units produced over time. Case in point: NYNEX, the telephone company, saved 500 tons of paper a year by reducing the size of its directories *one sixteenth of an inch* on all sides. Other cases in this chapter illustrate the same point.

Though cost reduction is the primary driver, other factors combine to make waste prevention in retail packaging a fairly continuous effort at many consumer product enterprises:

- New packaging materials and applications, especially the continuing development of plastics, offer opportunities to greatly reduce weight and cost.
- Security scanners at the exits of retail stores have moved the front line of defense against shoplifting away from the retail display rack, making it unnecessary to "overpackage" some products. CDs are an example.
- Cost analyses often call attention to packaging as a significant labor cost center, and that initiates conversations about packaging reduction.
- Brainstorms and innovations by employees are a constant source of improvement.

Hyde Manufacturing Co. illustrates the point about innovation. Hyde, located in Southbridge, Massachusetts, manufactures putty knives and other tools. For years the company had packaged its line of putty knives on hang-cards for retail display. One day a purchasing department employee wondered why it wouldn't be possible to eliminate the hang-card simply by drilling a hole in the putty knife handle and placing product information on an adhesive label affixed to the knife blade. A team including representatives of purchasing, marketing, consulting, manufacturing, and environmental departments worked up projections of cost and payback time. The numbers looked good, the project was implemented, and Hyde reduced its paperboard purchases by 8 tons a year, or about $40,000 of material. Equally impressive, the whole project took only 3 months from start to finish, indicating Hyde's commitment to waste prevention.

Teledyne Water Pik Co. made a packaging improvement that depended entirely on advances in plastic shrink-wrapping technology. Previously the company, located in Loveland, Colorado, had packaged its water filter cartridges in a paperboard carton for retail display. The brightly printed cartons did a good job of attracting attention. But they were expensive to manufacture, and assembly and loading were labor-intensive: eight employees were assigned to this task alone. With mass merchandisers among its main customers—KMart, WalMart, Builder's Square—Teledyne Water Pik needed ways to cut costs of warehousing, labor, and shipping. The new shrink-wrapper (actually a transparent sandwich of five separate layers of film, each with special properties) delivered on all counts, the company reported. Packaging material costs for this line of products were cut 70%, and labor costs reduced 87%. Warehousing requirements changed dramatically. Previously, to package a typical daily two-shift production run of about 60,000 filters required eight pallet loads of paperboard cartons. By comparison, a month's supply of shrink wrap on rolls fits on a single pallet.

These significant packaging savings were nicely capped by a very good reception in the marketplace and new outlets for Teledyne Water Pik's filter products. Interestingly, the switch to a transparent package is credited with improving sales. As a company representative explains, customers no longer have to open the package to see the product, speeding the decision to buy.

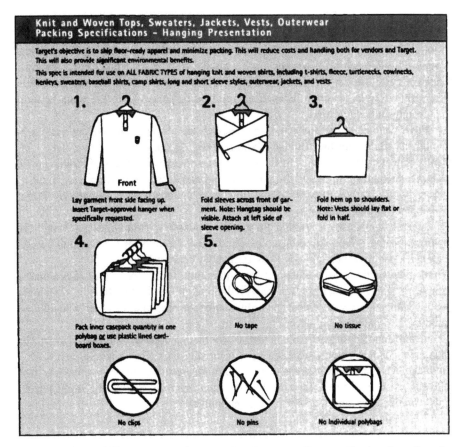

**Figure 1**  ON TARGET. Target explains to vendors its new packing specifications for softgoods. Goal is to "ship floor-ready apparel and minimize packing." Estimated total savings, $4.5 million. (Illustration: U.S. EPA.)

### CASE STUDY: Target Stores

Minneapolis, Minnesota

*Working with vendors to reduce unnecessary packaging of softlines merchandise, Target prevents 1.5 million pounds of excess packaging and saves $4.5 million in labor costs associated with unwrapping goods for retail display.*

Target Stores, headquartered in Minneapolis, is a national chain of 730 retail stores. When management became concerned that much of the merchandise shipped to Target included unnecessary packaging, a multidisciplinary team of 20 employees (representing the Environmental, Quality Assurance, Distribution, Operations, and Special Projects departments) was assembled to study the packaging of softlines merchandise, primarily clothing. After observing stockroom procedures in several stores, team members concluded that a great deal of labor was expended removing packaging from clothing prior to placement on the sales floor. For example, a shipment from a sweater maker contained 20 identical sweaters, and each was wrapped individually in a plastic bag. Each separate wrapper had to be removed before display of the merchandise. Clearly, the entire shipment of sweaters could have been safely transported from the manufacturer in a single outside wrapper.

When vendors were contacted, the Target team learned that vendors were supplying the packaging they thought Target wanted—a gap in communications, evidently. The Target team decided to experiment with test shipments of clothing with reduced packaging, to see if the goods would arrive unwrinkled and presentable. After test runs proved successful, packing specifications were developed by Target in cooperation with vendors.

"Target's objective is to ship floor-ready apparel and minimize packing," the instructions say. "This will reduce costs and handling both for vendors and Target. This will also provide significant environmental benefits."

The instruction sheet illustrates how garments should be folded. Specifically banned are tape, tissue, clips, pins, and individual polybags. In letters of agreement with vendors, Target reserves the right to impose fines on vendors who deliver merchandise not wrapped according to instructions.

In addition, Target has implemented a new hanger system. Clothing destined for display by hanger arrives from the vendor already on a plastic hanger and with a size tab. When hanging merchandise is sold, hangers are sent to a company that refurbishes them and sells them back to vendors for reuse. Previously, stores reused only good hangers and disposed of the remainder.

### Payback

In a single year, Target's program has reduced packaging by 1.5 million pounds and saved the company an estimated $4.5 million in labor costs associated with unwrapping merchandise. Vendors realized an estimated $3 million savings in reduced packaging material.

### Additional Waste Prevented

The hanger refurbishing program has eliminated disposal of broken hangers by stores. Now, the refurbisher grinds the unusable plastic hangers for recycling.

### CASE STUDY: Warner-Lambert Co.

Morris Plains, New Jersey

*Switching from a glass bottle with corrugated and paper overwrap to a plastic bottle without additional packaging, Warner-Lambert eliminates 20 million pounds per year of packaging for the Listerine mouthwash product, a 52% weight reduction.*

Warner-Lambert manufactures health care and consumer products. Listerine, which has been produced since 1906, is regulated by the U.S. Food and Drug Administration because of its claim to kill bacteria leading to gingivitis and plaque. Consumer complaints to WL about unnecessary packaging of Listerine led the company to examine the feasibility of making a change. The process required a sizeable capital investment, complex coordination among departments, long lead time, and patience. All the affected parties were involved: manufacturing, marketing, legal, purchasing, engineering, environmental, packaging technology. Among various factors considered were the following:

1. Redesign of the bottle in such a way as to deter competitors from copying the distinctive look of Listerine.
2. Development of a child-resistant cap closure that's easy for others to remove.
3. Shelf-life testing.
4. Drop-test proof.
5. Consumer acceptance—no decline in market share.
6. Safety. The old container presented the messy risk of a glass bottle filled with an alcohol-containing liquid. What were the risks associated with a new container?
7. New machinery. A single plant produces Listerine for the entire market.
8. Distribution, transportation—savings expected to accrue from reduced product weight.
9. Overall cost saving/payback.

The new Listerine bottle is manufactured of polyethylene terephthalate (PET) and has been well accepted by consumers, the company reports. Interestingly, WL tried to change its packaging this way in the early 1970s but was rebuffed. Focus group research at the time found that consumers felt such a packaging change signaled a cheapening of the product. Today, such waste prevention initiatives are applauded as evidence of a company's environmental sensitivity.

### Payback

WL will not reveal the overall cost of this packaging change. But the company says it is pleased with the change and expects to recover its investment in full. As one company official observes, a 52% reduction in packaging weight has to be significant.

WL has considered packaging changes for other products, with mixed results. For example, it was found that Caladryl, an anti-itch lotion for poison ivy, could not be packaged in PET (the most recycled plastic) because the lotion interacted with the plastic. Hence the bottle remained in a polypropylene container.

### CASE STUDY: Procter & Gamble

## Cincinnati, Ohio

*By redesigning the bottle, P&G reduces the Crisco cooking oil container by 30% and gains a number of other benefits.*

**Figure 2**  E-PACK. New Crisco container (right) uses 30% less material than original, saving 2.5 million pounds of plastic. It passes P&G E-test: efficient, economic, ergonomic, easy to use, environmentally OK.

When P&G sought a design change in the 32- and 48-ounce Crisco bottles, it found that a rectangular shape had more advantages than the old cylindrical shape. Thinner walls of polyethylene terephthalate (PET) were permissible, and the design passed the P&G "Total System" test:

1. Material—no breakage (PET was flexible enough to withstand drop-test failure).
2. Bottle supplier—able to manufacture the new design.
3. Design criteria—no dents (unappealing to customer).
4. Process—top-loading requirement (no bottle collapse during filling).
5. Trade standards—pallet and shipping (bottle shippers must fit pallet and be space-efficient).
6. Consumer acceptance—appearance, rigidity, shelf life, recyclability.

During 4 years of development and test marketing and careful analysis of cost, P&G partnered with its bottle supplier, Continental PET Technologies, and started producing the new "E-Pack" bottle (Efficiency, Economics, Ergonomics, Ease of use, Environmentally sound). The company summarizes benefits as follows:

1. Raw material requirements reduced 30% (a reduction of 2.5 million pounds of PET a year) from previous bottle and 48% less than competitors.
2. Weight reduced to 39 grams from 57 grams in the 48-ounce bottle; to 30 grams from 44 grams in the 32-ounce bottle. (P&G tried to reduce the weight of the smaller Crisco bottle to 26 grams, but the prototype did not pass the Total System test.)

3. Rectangular design permits more bottles to be shipped per cubic foot and per truck, saving shipping space and material.
4. Better space utilization with rectangular bottom, which reduces unused space to 12% vs. 22% for cylindrical bottom.
5. 10% reduction in corrugated cardboard—1.3 million pounds per year; reduced storage space.
6. Disposal benefits—bottle crushes flat; label glued in one spot only, easing removal; tamper-proof inner seal eliminated with introduction of tamper-proof outer cap, made of polypropylene, which is easily separated by density.

## Payback

P&G realized a 25–30% rate of return on investment and recovered all costs in about 3 years.

## Other Benefits

E-Pack also permits the following:

- Faster label assembly since one label has replaced two.
- Faster packing line changeover (reduced from hours to minutes) since the footprint of the larger Crisco bottle is identical to footprint of the smaller bottle, and case footprints are the same. Cylindrical bottles differed in width.
- Better customer acceptance since the grip is ergonomically optimal (less than 1-1/2 inches wide vs. 2 to 2-1/2 inches on competing brands—and on the old Crisco bottle).

"When the competition copies you, you know you've got a winner," a P&G staff member said.

## CASE STUDY: Sanyo-Verbatim CD Co., L.L.C.

### Richmond, Indiana

*Companies producing CDs eliminate 70% of the retail package by reducing the "long box" to a "jewel case" only.*

Before 1993, the standard CD retail package consisted of the plastic "jewel case" (the hinged, polycarbonate case with the CD inside) inside the "long box," a chipboard structure that doubled the length of the jewel case. The entire package was shrink wrapped. It measured 300 mm long by 150 mm wide. CDs were packaged this way because that's how retailers said they wanted them. The long box was considered desirable for: (1) Marketing—it provided advertising space, and (2) Security—it impeded shoplifters.

Phase-out of the long box began as an environmental initiative of the National Association of Record Merchandisers, whose members had been receiving growing criticism from record buyers about the over-packaging of CDs. The association

**Figure 3**   CD-PACK. Remember when CDs were packaged in the "long box"? Most of the
retail package went in the trash. Now, the "jewel case" is simply shrink wrapped.
Total reduction in materials cost: 70%.

adopted the position that after a certain date retailers would require all CDs to be
packaged simply in a "jewel case"—no long box.

Sanyo-Verbatim's experience in complying with this change probably was typical
of the industry. The company eliminated the machines that produced long boxes. It
also had to break down and repackage the inventory of CDs in long boxes—about
a million pieces. Special crews were hired for this 4-month task. And since the
industry standard bulk shipment of 30 CDs did not change, the original corrugated
cardboard containers designed to hold 30 long boxes became obsolete. The CD retail
package now measures 150 mm × 125 mm, or less than half the old long box.

### Payback

Compared to the old long box, the new CD package saves Sanyo-Verbatim about
$0.30 in packaging cost per CD. This extraordinary unit saving is the sum of the
following:

1. Elimination of long box production and labor.
2. Elimination of long box shrink wrap, a materials cost reduction of 70%.
3. From the reduction in volume, a 2–3 times increase in storage and shipping capacity.
4. 40% reduction in shipping weight per 30-count shipper—to 7 pounds from 12
   pounds.

Savings in package disposal cost also accrue at the receiving end by elimination
of more than half of the old CD package.

## Future

Manufacturers are looking at the possibility of packaging the CD in a cardboard case. Also under consideration is a method used abroad—a paper sleeve for the CD, with a cloth outer wrapper.

### CASE STUDY: Ben & Jerry's Homemade, Inc.

South Burlington, Vermont

*A two-piece retail package consisting of inner wrapper and outer box is replaced by a one-piece, plastic film wrapper, eliminating 11 million boxes and reducing waste by 165 tons.*

In 1987, Ben & Jerry's introduced the Peace Pop, a 3.7-ounce ice cream popsicle. The original packaging consisted of two layers—a poly-coated paper wrapper enclosing the frozen product, and an outer carton, used primarily because it provided space for advertising. Twenty-four individual popsicles were packed in a display container, and then in a corrugated shipper.

Ben Cohen and employees from marketing, manufacturing, and sales assembled to evaluate packaging of the Peace Pop, taking into account these factors:

1. Customer impact
2. Operational impact
3. Price and availability
4. Environmental attributes

Alternative new wrappers ranging from paper to plastic film were tested. Recyclable paper was evaluated but failed to pass tests simulating the strains of shipping and market display. A heat-sealed, oriented polypropylene wrapper often used in the ice cream market was not strong enough for the Ben & Jerry's 3.7-ounce product.

The wrapper finally selected was a 48-gauge, 1-1/2 mil opaque sheet of polyester and Surlyn ionomer. Like many composite plastics, the new Peace Pop wrapper is generally not recyclable.

### Payback

Ben & Jerry's packaging change was not motivated by cost reduction but source reduction. But in fact, savings resulting from elimination of the outer carton were balanced by upgrading to a more expensive single wrapper. Cost-wise, it was a wash.

### Benefits

The new Peace Pop wrapper weighs less than the old two-piece package and requires less energy to produce. Based on projected annual sales of the Peace Pop, Ben & Jerry's calculates that this packaging change eliminates the disposal of

**Figure 4**   BAN THE BOX. On its new Peace Pop wrapper, Ben & Jerry's declares, "Source reduction is the best environmental commitment anyone can make." Eliminating outer box reduced paper needs by 165 tons.

11 million popsicle boxes weighing a total 165 tons. In addition, packing at the ice cream factory has been simplified by elimination of the automated boxer previously required to pack popsicle cartons. Now, the wrapped popsicle is fed directly into a shipper which serves double duty as a display case. Ben & Jerry's also saves by

1. Eliminating the old outer shipping carton.
2. Reducing shipping costs per unit because of lighter unit weight.

# CHAPTER 9

# Advancing Toward Zero Waste:
# Back to the Future

In 1926, Henry Ford observed, "Waste is not something which comes *after* the fact. Picking up and reclaiming the scrap left over after production is a public service. But *planning* so that there will *be* no scrap is a higher public service."

Ford, who was obsessed with reducing the waste of material and time, may have had a particular example in mind. He and his engineers had noticed that the process of stamping out door panels for the Model T produced a large amount of scrap steel. Of course, the scrap could be picked up and returned to the melting pot, but wasn't there some way to add value to this material? Ford and his engineers found that there was—by using the door scrap as raw material for stamping out fan-drive pulleys. This little bit of waste prevention reduced Ford's steel needs by 300,000 pounds a year.

In 1996, Ron Simmerman, a 31-year veteran of the General Motors Powertrain plant in Bedford, Indiana, guides a visitor through the machine maintenance section of the big GM factory. This is where they make aluminum piston heads, transmission casings, and engine blocks for Cadillacs and other models. We stand at the parts and tool rack. Nearby, millwrights are busy at work. Simmerman pulls a small plastic box full of electrical connectors from the rack. When the box is empty, he explains, it's returned to central stores for refilling. The empty box itself serves as notice to send a full box of parts. That's the *kanban* (just-in-time) system, and among other things it means better productivity by eliminating the waste of high-priced union labor waiting in line at the supply window.

Let's back up a step. Originally, the plastic parts boxes were made of corrugated cardboard, used once and tossed. That created a housekeeping problem, a messy accumulation of empty boxes. The Synchronous Manufacturing Team at Bedford looked at the situation and recommended switching to the reusable plastic containers, and it was done. But now Simmerman has something even better in mind, and it would have been unthinkable only a few years ago (surely not in Henry Ford's time). Instead of using GM employees to tend this internal parts supply system, why not contract with the outside company that supplies parts to GM to monitor inventories and deliver parts directly to the parts rack deep inside the plant, bypassing central stores and eliminating some paperwork, too?

This train of creative thinking at GM Powertrain illustrates what can happen when people do not settle for the first answer to a simple question such as, How can we avoid the housekeeping mess created by a pile of empty cardboard boxes? And it also reflects the practical application of certain theoretical pronouncements about waste.

Definitions of waste emerging from Germany are especially interesting. The German Recycling and Waste Management Act of 1996, for example, separates all goods circulating in an economy into just two categories: products and waste. But it defines waste in a way that makes it absolutely inseparable from the production process. The net effect is to attach the responsibility for managing waste to the manufacturer from the very inception of the product life-cycle. This is also known as the Polluter Pays Principle, an idea that has spread from Germany to other European nations and is beginning to influence the behavior of U.S. companies doing business abroad. Let's examine why.

## THE GERMAN INITIATIVE

The German law cited above is rooted in a 1991 enactment of the German parliament, "An Ordinance on Avoidance of Packaging Waste" (*Verpackungsverord-nung,* or *VerpackVO*). The underlying concept is that manufacturers are responsible for taking back their packaging, with the hope and expectation that this will lead to designing packaging for low toxicity and pollution, for waste minimization, for reuse or recyclability, and for conservation of materials. Germany is the first nation to enact such a law. The ordinance covers most kinds of packaging in the marketplace. It defines three categories:

1. *Transport packaging.* This packaging "protects goods in transit…. In general, it is removed by the distributor and does not get as far as to the final consumer." Continuing to quote from the ordinance, transport packaging means "drums, containers, crates, sacks including pallets, cardboard boxes, foamed packaging materials, shrink wrapping, and similar coverings which are component parts of transport packaging and which serve to protect the goods from damage during transport from the manufacturer to the distributor or are used for reasons of transport safety."
2. *Secondary packaging.* This includes "blister packaging, film, cardboard boxes or similar coverings which, as an additional layer of packaging around sales packaging, are intended to allow goods to be sold on a self-service basis, or to hinder or prevent the possibility of theft, or principally for advertising purposes."
3. *Sales packaging.* Such packaging is in use "until the goods are consumed and/or put into use by the final consumer and does not lose its protective function before then. The final consumer usually acquires packaged goods in the shop and takes them home with him/her. Final consumers can equally be either private, commercial, or industrial final consumers."

Obviously, our interest in the German law has mainly to do with transport packaging, and later in this chapter, the IBM case shows how the German enactment directly affects packaging decisions by U.S. manufacturers. In broad terms, packaging arriving in Germany is assessed a fee based on the kind of material and its weight, generally reflecting the cost to reuse or recycle rather than dispose of the material. In one fee schedule, for example, paper and cardboard are charged $0.09 per pound; plastic, $0.72. It is easy to imagine how a 63-cent difference in disposal cost per pound might influence the selection of packaging material!

Concerning retail packaging, Germany's packaging fees have led some U.S.-based manufacturers to redesign containers and benefit from very significant reductions in cost. Procter & Gamble, for example, eliminated the outer carton of a denture cleaning product and cut 25 tons from its annual purchase of cardboard. P&G also redesigned a detergent container, reducing plastic consumption for one product line by 45%.

Most other European countries have adopted take-back laws and tariffs similar to Germany's. Could such laws and regulations be enacted in the U.S.? For a number of reasons, that seems unlikely. Irrespective of the cost, disposal is not viewed as a significant problem in the U.S. by comparison to Europe, with its dense population. More to the point, federal rulemaking is extremely unpopular here. But Germany's initiatives undoubtedly will influence packaging directions in the United States simply because of the importance of U.S. export trade and the huge volume of goods we ship abroad.

Within their own borders, European countries also are demonstrating wider use of returnable transport packaging than is evident in the U.S. In selected applications, the food industry in the United Kingdom is rapidly replacing corrugated cartons with returnable plastic crates. Crates have been used for years to deliver dairy products and baked goods; now the returnable containers are carrying produce, meat, and other chilled products. In 1996, one supermarket chain was transporting 800,000 filled plastic crates of food items per week, replacing about 42 million corrugated cartons per year. Forecasts about the use of crates are bullish. In the U.K., where 9.5 million were in circulation in 1992, 20 million were in use by 1996, and 28 million were projected for 2000. By that year, the U.K. is forecast to be using more crates than Germany, where the movement began.

Why are reusable crates so rapidly invading the territory so long occupied by corrugated fiberboard? The basic reason is to improve operational efficiency, observers say. But surely this is a complicated proof, weighing the ease of baling and recycling corrugated, a well-understood business routine, against the new logistics of backhauling, washing, and reusing plastic totes. It may be more productive to think of the attention to waste prevention now occurring in transport packaging and all other aspects of production and distribution as part of a growing sense of product stewardship that spans the entire product life cycle. To many U.S. companies, this was a foreign concept only a short time ago. Today, life-cycle design and assessment are being driven by good business sense as well as environmental stewardship.

## GETTING STARTED

Launching a waste prevention program may seem a daunting enterprise. It needn't be—companies have been going about this business for decades. The main difference today is that companies are learning from one another. The largest organized effort among U.S. companies to prevent waste is the U.S. EPA's WasteWi$e program. Since its founding in 1994, WasteWi$e has enrolled some 600 companies. Their combined efforts are impressive. An EPA report says partners in the voluntary program eliminated 453,000 tons of materials through waste prevention activities during 1996, a 30% increase over the previous year. For the group as a whole, avoided disposal fees were $15.4 million in 1996; avoided paper purchase costs totalled $64.5 million.

Partners in the program are asked to make a 3-year commitment to carrying out specific waste prevention activities, beginning by examining operating and purchasing practices to identify targets for action. The objective within the first 6 months is to select five practical and measurable goals tailored to their needs. WasteWi$e offers several kinds of assistance, including technical resources, waste audit programs, toll-free helplines, a partner network, and recognition. The main thing asked in return is an annual report of progress so that EPA can track overall results. Further information is available in several forms: by telephone to 800-EPA-WISE (372-9473); by e-mail to ww@cais.net; from the homepage, www.epa.gov/wastewise; and by mail to WasteWi$e Program, U.S. Environmental Protection Agency (5306W), 401 M St., S.W., Washington, DC 20460.

Over the years, many good ideas about waste prevention have come from the state of Washington. Recently, the Snohomish County Solid Waste Management Division, in Everett, initiated a Packaging Waste Prevention Project for area businesses—free advice and assistance. Two dozen businesses participated and 13 actually made changes in packaging. (One of them, Alpine Windows, is presented as a case study in Chapter 1.) By trimming away unnecessary materials and becoming more efficient, the 13 businesses reduced their costs a total of $440,000 a year. What's interesting is that all these enterprises were, to one degree or another, skeptical about the value of time spent examining new opportunities for waste prevention. "We've already done that," they said. Results of the Snohomish County project demonstrated that, even at companies with the highest sensitivity to waste avoidance, it doesn't hurt to take another look around.

### CASE STUDY: Stanpac

Smithville, Ontario

*Refillable glass bottles provide a significant cost advantage to small regional dairies.*

Stanpac designs and manufactures closures and application equipment for the dairy, juice, beverage, pharmaceutical, and healthcare packaging industries. For dairies using refillable containers, Stanpac is a major producer of equipment and a source of technical information.

Originally a producer of milk bottle caps, Stanpac was forced to diversify and began to manufacture closures for other products when the milk container market shifted from glass refillables to cartons. Glass milk bottles were practically gone—but not forgotten. In the late 1980s, local dairies that wanted to bottle their own milk began to emerge, creating a very small but measurable increase in the demand for refillable containers. These new-generation dairies wanted a container that would make their product stand out, allow economical distribution, and create a premium product image. At that time there was only one supplier of refillable glass bottles. Recognizing the beginning of a trend, Stanpac purchased molds to produce glass milk bottles and now markets them in 8-, 16.9-, 32-, and 64-ounce sizes plus a 1-liter bottle. Most sizes are closed with a paperboard disk crimped around the mouth.

### Payback

In 1996, milk in refillable containers constituted slightly more than 1% of all liquid milk shipments. Volume is increasing at the rate of 15% annually. Although there is no unusual economic benefit to Stanpac from producing glass milk bottles, the company's action has enhanced its image and promoted a positive public perception of the container. Stanpac anticipates that as the market for refillable glass bottles grows, it will have the opportunity to expand production of related products, such as closures and divided cases.

### Additional Waste Prevented

The reuse of glass containers saves between 80% and 90% of the energy required to produce virgin glass. By comparison, recycled glass saves between 10% and 15% of the energy required to manufacture new glass. Refillers of glass bottles generally rely on backhauling of empty bottles collected from users during regular route deliveries; thus, no additional vehicle trips are necessary.

Typically, a refillable glass container is reused an average of 25 times. According to the Recycling Association of Oakland, California, a bottle refilled 25 times will use 95% less glass and 90% less energy than the total process of producing 25 bottles in closed-loop recycling. This is true even when factoring in the total cost of inspecting, sterilizing, and refilling reusable glass bottles. A refillable glass container system also has a higher jobs-to-volume ratio than a recycled glass system.

### CASE STUDY: IBM Corp.

Charlotte, North Carolina

*Shipping costs into the European market are reduced by replacing expanded polystyrene packaging materials with corrugated paper pads.*

An updated packaging system was needed for an updated IBM product, the 4772 banking printer. Previous packaging used expanded polystyrene (EPS) cushioning and corrugated components, resulting in a large, bulky package to provide adequate shock protection during transit. IBM packaging engineers wanted to take an environmentally friendly approach. Their goals were to eliminate commingled packaging

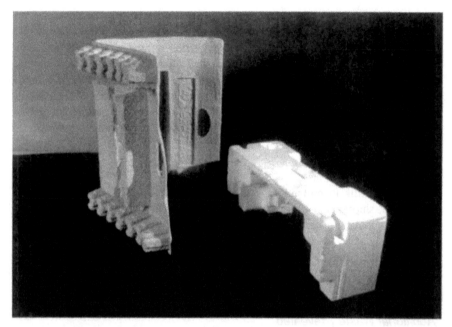

**Figure 1**  EXPORT-PACK. European countries charge manufacturers for disposal of import packaging. Tax on the molded paper container at left is $0.09 per pound; polystyrene on right, $0.72. Which do you specify?

(EPS plus corrugated), provide equal or greater product protection, use more readily recycled materials, and reduce the overall cost of the process.

IBM selected a product called Corrupads to accomplish this purpose. Corrupads are corrugated cushioning pads produced from 100% post-consumer waste paper. The multiple layers of the pads absorb impact and minimize the rebound characteristic of resilient foam materials. Thus, shock levels during handling and transit are reduced. The new packaging system exceeded IBM's test specs in this case, a drop height of 30 inches on a total of 8 sides, plus sine sweep vibration and random vibration.

## Payback

There was an immediate saving of more than $20,000, because the new material does not require the tooling necessary to produce EPS components. Since less packaging was required to produce the same cushioning effect of polystyrene or polyethylene foam, total package size could be reduced. This size reduction led to a 10% reduction in shipping expense and a 25% reduction in storage space requirements, including pallet load maximization. Additional savings were realized in shipments to Germany, for example, where "Green Dot" legislation requires the assessment of fees for packaging material based on the ease with which it can be recycled. Packaging in the paper category, which includes the Corrupad, is charged $0.09 a pound. By contrast, plastics such as EPS are charged $0.72 per pound. Overall, IBM reduced shipping–handling costs for the new banking printer by 70%.

*Additional Waste Prevented*

Because the exterior corrugated box—the shipper—could be smaller, less corrugated was required. The smaller package meant more packages could be loaded on a pallet, and that reduced the need for pallets by 25%. The Corrupad product also provided a market for post-consumer paper. The many benefits of converting packaging materials for the new banking printer persuaded IBM to do the same thing with an earlier model of printer.

## CASE STUDY: Herman Miller, Inc.

## Zeeland, Michigan

*The Avian chair from Herman Miller illustrates how a company can build the abstract concept of "lowest possible ecological impact" into a tangible, complex, and very attractive piece of office furniture.*

Herman Miller believes that longer-lived products translate into better stewardship of the earth's resources. In addition, the company believes that more durable products are easier to refurbish for extended use. In 1994, the company formed Earth Friendly Design Task Force and assigned it to infuse the company's design process with environmental values. Life-cycle analysis has become a major focus of the task force. Such an analysis considers all the materials that go into a product in light of recycled content and recyclability. The task force also examines new products to determine how they can be designed for disassembly, facilitating reuse and recycling at the end of the product's life.

The Avian chair illustrates all the foregoing. For example, gas-assist injection molding is used to manufacture a hollow frame (like the bones of a bird, hence the chair's name). This design feature preserves structural strength while reducing materials and weight. The frame requires no paint or other finish, conserving materials. All components are recyclable and none are ozone-depleting. The chair is shipped partially assembled, reducing unit packing size, increasing cube utilization, and requiring less transport energy per given quantity of chairs. A schematic, modeled after work done by the American Society of Testing Materials, is imbedded on the bottom side of the seat (see illustration) to identify all parts by kind of material.

*Payback*

Herman Miller requires any system change to pay for itself in less than a year. Although the company has not released any cost information about the Avian chair, it is assumed this project met guidelines.

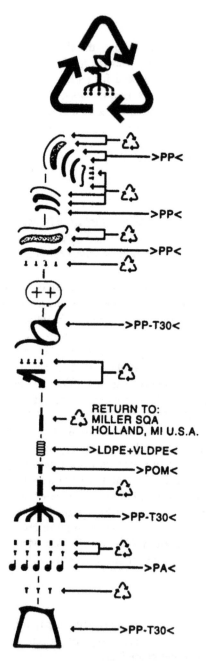

>PP<

>PP<

>PP<

>PP-T30<

RETURN TO:
MILLER SQA
HOLLAND, MI U.S.A.

>LDPE+VLDPE<

>POM<

>PP-T30<

>PA<

>PP-T30<

**Figure 2**  TOTAL RECOVERY. First, Herman Miller minimized raw materials in the Avian™ office chair. Then it molded reuse–recovery instructions into the seat bottom, like a friendly message to a future civilization.

# Packaging Instructions to
# General Motors Assembly Plant Suppliers

The automobile industry was one of the first to demonstrate the effectiveness of waste prevention initiatives in transport packaging. Appendix A presents in its entirety the instructions about transport packaging that General Motors distributed to all of its suppliers in 1994. It concerns pallets, pallet cartons, the use of dissimilar materials, dunnage, strapping, unitizing, and other topics. The GM "Environmental Packaging Addendum," which is reproduced with permission of General Motors, presents an excellent model for other companies interested in adopting similar standards.

## North American Operations

September 1, 1994

TO:   All General Motors Assembly Plant Suppliers

# Environmental Packaging Addendum

The following packaging addendum to the GM 1738 Packaging and Identification Requirements for Production Parts supports the General Motors "WE CARE" Program (Waste Elimination and Cost Awareness Rewards Everyone). The changes are based upon sound engineering principles, customer input and the latest in packaging technology.

### Supplier Environmental Packaging Requirements

(1)  Corrugated pallets are required for all loads less than 500 pounds. Any instances of excessive cost penalties must be reviewed satisfactorily with NAO Containerization.

(2)  Pallet cartons must be constructed with a "breakaway" feature or other method to be easily separated from the shipping pallet.

(3)  Wood is not to be stapled to the corners, walls or the top of any carton.

(4)  Foam glued to corrugated and other dissimilar packaging materials bonded together is not allowed. All foams and foam sheeting are discouraged.

(5)  Expanded polystyrene (EPS) must not be used as an expendable packaging material.

(6)  Non-metallic strapping must be color-coded based on Automotive Industry Action Group (AIAG) standards, and closure must be by friction sealing.

(7)  Unitizing adhesives are strongly encouraged as a replacement for stretch wrap or strapping.

(8)  Use clear (non-tinted), low-density polyethylene stretch film (LDPE).

(9)  Identify rigid, molded plastic with the proper resin code.

(10)  Wasteful, excessive and non-recyclable packaging will not be acceptable.

(Items 1 through 10 are expounded upon in the attachments)

NAO/CSG/2856                    ( 9/1/94 )

# #I PALLETS

**Background:**

The disposal or recycling of wood pallets continues to be the greatest challenge for our plants to manage. Our plants will continue to receive some wood pallets, and efforts are being made to reuse or recycle as many as possible, but to reduce this impact:

**CORRUGATED PALLETS ARE REQUIRED FOR ALL LOADS LESS THAN 500 POUNDS.** Any instance of excessive cost penalties must be reviewed satisfactorily with NAO Containerization.

Corrugated pallets with loads above 500 pounds for any size pallet are encouraged but optional. Products currently shipped on returnable pallets will continue to be shipped that way.

**We will continue to accept used wood pallets that meet GM 1738 requirements.** The responsibility for the quality and performance rests with the supplier. To assist recycling, wood pallets must have the pallet size marked on the runner in 1" minimum characters.

It is recommended that implementation of corrugated pallets begins soon to allow for selection of the best pallet, to phase in the concept, to make any necessary test shipments, and to receive proper feedback. Upon request, the NAO Containerization Group can provide suggestions for sources, the transition into corrugated pallets, and assist with test shipment coordination.

**Items to consider in selection of corrugated pallets:**

(1) Continuing improvements in pallet design, more suppliers, and increasingly competitive prices of corrugated pallets suggest periodic review.

(2) Structural members of the pallet should be compatible with your carton by supporting the edge and corners.

(3) A solid corrugated deck is desired.

(4) If paper fiber cores are used for load-bearing members, use no more than four with a maximum thickness of 1/4".

(5) Recyclability of pallet (100% corrugated preferred) is required.

(6) Identification of manufacturer and/or pallet name printed on the pallet runner is REQUIRED.

(7) Fork opening must be compatible with GM 1738 requirements. (2" x 8", 18" centers, 26" minimum inside dimension).

(8) Strength to stack full three-high in storage or to a height of 10.5', whichever is greater.

(9) Two-way entry is permitted for runner lengths up to 48". Four-way entry may be required on a notified basis.

NOTE: Supplier is responsible for packaging performance. Test ship if necessary.

NAO/CSG/2856

9/1/94

# #2 PALLET CARTONS

**Background:**

Corrugated flanged tubes fastened to a pallet (referred to as pallet cartons) are generally used for bulk shipments. Bottom flanges are fastened to the wood pallet around the perimeter to secure the two units. This has been an effective method of shipping bulk items, but our assembly plants' ability to separate the two materials for recycling is impaired by the current fastening process. Improvements have been made in pallet carton design that allow quick and effective disassembly of the corrugated sleeve from the pallet.

**ALL PALLET CARTONS MUST BE CONSTRUCTED WITH A "BREAKAWAY" FEATURE OR OTHER METHOD TO ALLOW EASY SEPARATION FROM THE SHIPPING PALLET.**

This concept involves simple die-cut areas (generally "U-shaped") perforated properly and spaced along the lower flanges with the open end of the "U" facing the center of the pallet. The corrugated-to-pallet staples are placed within these areas. Four hand openings, die-cut flaps in the sleeve, one per side, allow for lifting the sleeve from the pallet when disassembly is required. Printing on two panels is required to identify this feature. Two-inch-high bold type, for example states "breakaway," "fast break," "E-Z Pop," etc.

NOTE:   Consistent, correct assembly of the carton (i.e., stapling inside the die cuts and using corner "kick-ins" to contain wood supports) requires proper explanation/training of the carton assembler.

NAG/CSG/2856                                                              9/1/94

# #3 WOOD CORNER SUPPORTS

**Background:**

All corrugated cartons are to be recycled after use. Any contaminant such as wood that is fastened to the corrugated material requires intensive and costly labor to separate.

**Wood is not to be stapled to the corners, walls or the top of a carton.**

**ALTERNATIVES:**

(1) Eliminate the wood if possible. The carton strength may be currently acceptable.

(2) Replace the wood with formed paper corner structures, or corrugated supports. These paper corners may be stapled in place, since they can be recycled with the sleeve. The use of "angle board" that has a white clay coating and non-water-soluble adhesive is not allowed. The cross-sectional area allows virtually no surface to support a load.

(3) As a last resort, wood may be used as a corner support but must NOT be stapled to the corrugated supports. Other methods of holding the wood in place must be used, allowing ease of wood removal. Corner "kick-ins" or corrugated pockets will suffice. Metal staples are acceptable for carton closure or to staple paper-based supports to the carton.

NAO/CSG/2856

9/1/94

# #4, 5 FOAM DUNNAGE & DISSIMILAR MATERIALS

**Background:**

Foam dunnage protects parts from abrasion, part-to-part contact, and supports items in transit. Currently, we have no economical solutions for recycling foam, which causes us to landfill the material or return the dunnage to the suppliers at their expense. The use of foam or foam sheeting as an expendable material is discouraged.

**THE FOLLOWING CANNOT BE USED AS EXPENDABLE PACKAGING:**

**FOAM GLUED TO CORRUGATED AND ANY OTHER DISSIMILAR MATERIALS BONDED TOGETHER.**

**EXPANDED POLYSTYRENE (EPS) "STYROFOAM"**

**ALTERNATIVES:**

(1) Eliminate the foam or replace it with other easily recycled materials such as die-cut corrugated or molded Kraft paper pulp.

(2) Sensitive part surfaces placed near corrugate may require recyclable polyethylene film or bags, or repulpable coatings on the corrugate surface. Wax coatings are not permitted, and foam sheeting is undesirable.

(3) Mechanically attach the foam to corrugated or other dissimilar materials to allow for ease of disassembly and subsequent recycling of the corrugated.

(4) Reduce the amount of foam to a minimum.

Foam is a relatively expensive material for one-way shipment of components. Our suppliers are continuing to use the above alternatives to reduce their packaging costs.

NAO/CSG/7856

9/1/94

# #6 NON-METALLIC STRAPPING

**Background:**

Plastic (non-metallic) strapping is available in various materials that satisfy requirements of tensile strength, elongation and recovery. Additionally, strapping is available in many colors from many suppliers.

With our need to separate and recycle all packaging, it is not economical to analyze or test all strapping to determine its composition. In addition, mixed colors have little or no market value for recycling strapping.

**STRAPPING COLOR MUST BE STANDARDIZED USING THE AIAG STANDARD OF:**

POLYESTER - TRANSLUCENT GREEN
POLYPROPYLENE - TRANSLUCENT CLEAR

**The GM 1738 recommends polyester strapping due to its strength and recovery proper-ties.** Use of any other strapping requires approval by a representative of the NAO Container-ization Group.

Nylon has no color standard applied to it. We know of no process or program to effectively recycle nylon; its cost is excessive, and its performance can be matched with polyester strap-ping.

**FRICTION SEALING IS REQUIRED OF NON-METALLIC STRAPPING.**

Metal clips or buckles are prohibited. Elimination of the metal clip reduces cost and effectively allows recycling without the potential of metal contamination.

The sealing tools to perform this function are powered by either compressed air, electricity or battery.

NAOPC5G/2656

9/1/94

# #7, 8 PALLET UNITIZING METHODS

Plastic strapping and plastic stretch wrap have been the acceptable method of securing cartons to a pallet. Our assembly plants are working to recycle all packaging materials; strapping and stretch wrap included.

**As a supplier, you can assist our efforts by:**

**Strapping** - *Incorporate the standardized colors and strap-sealing method required in the strapping section.*

**Stretch film** - *Use clear film; the absence of color maximizes recycling potential. Specify LDPE (low-density Polyethylene) film.* LDPE films are the most common, highest performing types of film and should be used. PVC film is not to be used. Wrap the palletized unit with only the required amount of film, minimizing waste for all. Your film supplier or equipment supplier can assist you with using stretch film most effectively.

**Unitizing adhesives** - *This special adhesive is the best environmental option, requiring no strapping or stretch film to be collected, processed or recycled.* Recycling of the carton is unaffected. Many suppliers are using this concept, and it is widely accepted by our plants. Supplier benefits are:

(1)  Wrap time eliminated.
(2)  Equipment/floor-space savings.
(3)  Load integrity, appearance.
(4)  Inspect/repack ease.
(5)  Unit load increased. Nothing stretches/moves. Possibility for corrugated reduction.

# #9 SPI PLASTIC RESIN CODES

**Background:**

To facilitate the recycling of a product, its identity must be known. There are numerous types of plastics used for automotive packaging which require a simple method of identification. NAO will require the SPI (Society of Plastics Industry) coding, the same as on retail packaging, familiar to all. The SPI code chart is shown below.

All vacuum-formed and injection-molded plastic packaging material must be identified by this code.

NOTE: Plastic components that are assembled to the vehicle are to be identified with the proper SAE code to facilitate recycling. Packaging will be marked with SPI codes.

## KEY:          SPI CODES

PET
*Polyethylene Terephthalate*

HDPE
*High-Density Polyethylene*

V
*Vinyl/Polyvinyl Chloride*

LDPE
*Low-Density Polyethylene*

PP
*Polypropylene*

PS
*Polystyrene*

OTHER
*All Other Resins and Layered Multi-Material MDPE - PE/PP*

9/1/94

# #10 WASTEFUL, EXCESSIVE, OR NON-RECYCLABLE MATERIAL

**Background:**

Packaging is required to serve many needs; part protection, transportation effectiveness, synchronous manufacturing, and ergonomic and environmental concerns to name a few. Proposed and impending state and federal legislation is prohibiting wasteful and/or excessive packaging. The challenge then is to meet these requirements with the amount and degree of packaging necessary and no more. **Overpackaging and wasteful "just-in-case" packaging is undesirable for both the supplier and the user.** Each NAO supplier is expected to identify and correct such packaging on an ongoing basis.

With reduction or elimination as the first priority, the hierarchy of waste elimination is:

### REDUCE          REUSE          RECYCLE

To list every example of wasteful, excessive or non-recyclable packaging would be too extensive. We have identified a few examples that have been significant problems at our plants:

- Cartons partially filled

- Oversized foam, plastic or corrugated dunnage

- Microcellular foam wrap and bubble wrap

- Plastic protective covers, caps, plugs, paint masks or spacers required in the manufacturing process but not required as a protective shipping device

- Corrugated carton test strength that far exceeds requirements

**Non-recyclable packaging** is that which has no available or economical system in place to process an item. Wax-coated corrugated is a prime example of this type of packaging.

**Waxed- or plastic-coated paper is prohibited** because it contaminates the recycling process.

Non-kraft corrugated has no recycle value and; therefore, is unacceptable. Recycling centers will not accept it; therefore, we will not accept it.

**Plastic plugs, caps, and protectors** are extremely difficult to recycle due to oil and paint contamination, colors, uncertainty of resin type, and transportation costs. Every effort should be made to eliminate the plastic. If it cannot be eliminated, other changes can be made to assist the plants' recycling efforts:

(1) Mold the Society of Plastics Code (#1-#7) into the part. When elimination is not possible, these codes will allow for effective recycling.

(2) Clear LDPE plastics are preferred and can be effectively recycled.

(3) Ship plastics uncontaminated with paints and lubricants.

(4) Replace the plastic with a paper substitute.

**Any plastic cap, plug, spacer, etc., if not required for packaging or shipping protection, must be removed prior to shipment to the assembly plant.**

NAO/CSG/2856                                                              9/1/94

# Index